INTERDISCIPLINARY MATHEMATICS

VOLUME XII

THE GEOMETRY OF NON-LINEAR DIFFERENTIAL EQUATIONS,
BÄCKLUND TRANSFORMATIONS, AND SOLITONS

PART A

ROBERT HERMANN

MATH SCI PRESS
53 Jordan Road
Brookline, Ma. 02146 (USA)

Copyright © 1976 by Robert Hermann
All rights reserved

Library of Congress Catalog Card No.: 76-17201
ISBN 0-915692-16-3

MATH SCI PRESS
53 Jordan Road
Brookline, Ma. 02146

Printed in the United States of America

Preface

In this Volume, I have embarked on a vast project--the exposition and development of the 19-th century geometric theory of differential equations, particularly in terms of applications to engineering and physics.

Now, those familiar with other branches of mathematics which were extensively worked on in the 19-th century (e.g., complex variables or algebra) might think this is a relatively simple matter of historical research. In fact, it is primarily a matter of new mathematical work, since so much of the material has been completely lost to present day science. My task is to reconstruct it, much in the way that a paleontologist reconstructs a dinosaur from a bone in the foot.

I have used as guides three major works. First, the articles in the Enzyklopëdie der Mathematischen Wissenchaften (or the French version, Encyclopédie des Sciences Mathematiques), particularly the articles by Von Weber and Vessiot. Second, E. Cartan's articles and books on "exterior differential systems". (Much of his work in this area was, in fact, oriented towards unifying the work of his predecessors by developing it in the context of his own way of looking at differential geometry.) Third, Darboux' "Theorie des Surfaces" which is really much more about the theory of non-linear partial differential equations than it is about "surfaces" in the usual sense.

Although I have been interested in this topic for a long time (for example, see my article "E. Cartan's Geometric Theory of Partial Differential Equations"), the factor compelling me to embark on this almost impossible task is a striking recent development in mathematical physics--the revival of material that was done in the classical period in terms of what is now called the Theory of Non-Linear Waves. (See Whitham [1].) In particular, recent work on solitons is so strikingly a revival of ideas of the 19-th century that it cannot be a coincidence--in fact, the secret to understanding the nature of elementary particles may have its mathematical roots in this work. I believe that developing this material in the context of modern mathematics will make fundamental contributions to mathematical engineering and physics, as well as stimulating pure mathematical research itself towards working on material that is of far greater intrinsic interest and importance than many of the topics that are fashionable today!

This first volume contains two major topics. First, a reworking of certain general material in Von Weber's encylopedia article, and then a much more specific topic, my own version of very recent work by two mathematical physicists, Frank Estabrook and Hugo Wahlquist, on a differential geometric setting for the recent work on non-linear waves. What is particularly significant is that they have developed a general setting for two key geometric ideas--"conservation laws" and "prolongation", and shown the very natural way that Cartan's formulation in terms of exterior differential systems leads to these notions.

What has this to do with elementary particles? First, I must say that I believe that 99.98...% of the modern work by physicists on what might be called Fundamental Physics, particularly on finding an appropriate mathematical setting for understanding the experimental elementary particle phenomena, is wrong-headed and even perverse. It is all basically motivated by the wish to take the <u>linear</u> equations and quantum theory with which they are familiar and modify it in some way to account for the observed phenomena, which are strikingly non-linear. They expect mathematics to give them a magical hocus-pocus to do this, but I strongly believe (often by study of my own, part of which has gone into my earlier books) that what they want to do is impossible and that the whole business must be rethought. A truly non-linear theory will be completely different than what is now sold to us as Quantum Field and Elementary Particle Theory.

Why do I want to go back to 19-th century ideas? I believe that this was the period when differential equations were thought about <u>geometrically</u>, and most geometrically-inspired differential equation theory is intrinsically non-linear. If 19-th century geometry were developed along the same lines as modern elementary particle physics, the only surfaces to have been investigated in Darboux' treatise would be planes and those which are "perturbations" of planes!

Notice that two of the most striking and conceptually successful recent ideas in elementary particle physics--gauge (i.e., Yang-Mills) fields and solitons--are closely linked to geometry. In fact, I will develop here Estabrook and Wahlquist's insight into the mathematical nature of solitons in a form that will provide a link between them, involving the theory of connections in fiber bundles. I believe that quantum mechanics for such geometrically inspired objects must be thought out completely anew. I am always horrified to see physicists proudly butchering beautiful mathematical concepts by trying to fit them into the standard ideas of perturbation-Feynman diagram-renormalization-quantum field theory. (Unfortunately, "axiomatic-constructive" quantum field theory is no better, since geometric-Lie group concepts play such a small role in the underlying intuition and esthetics. I just do not understand why a geometric object

PREFACE

like a connection should be "quantized" as an "operator-valued distribution", and I think it is rather mindless to set this up as an "axiom".)

Like the small mammals that developed while the dinosaurs reigned, I see certain modest theoretical ideas that are promising for a correct mathematical understanding of the elementary particle phenomena. Evidently, I believe that the soliton ideas are of this type. Here one encounters purely and superbly non-linear phenomena, that seems to offer a new approach to the traditional "wave-particle" conundrum. Unfortunately, it is unlikely that the striking particle-like properties that have been found in equations of one space dimension will extend to three dimensions. However, something must generalize--perhaps the topological properties mentioned below. Mathematically, these solitons are closely linked to Bäcklund transformations--surely, they must generalize, and they might even be the key to understanding how Lie groups enter into elementary particle phenomena. In fact, in the last chapter, I present some work lifted from Darboux that shows, for certain linear partial differential equations, how Bäcklund transformations appear as gauge transformations. It is interesting that two basic field equations--the Klein-Gordon and Sine-Gordon equations--appear in a very natural way. I hope this will convince the skeptical reader that there are many goodies buried in the 19-th century literature that have been completely forgotten, but that are very relevant to today's problems in physics. (Another example is a fact I have recently discovered about a relation between classical invariant theory and quantum mechanics--see my paper "The Poisson-Moyal Bracket,...")

Of course, the Bäcklund transformations are a facinating topic in themselves, that could well absorb all of my energies for many years. Apparently nothing has been done about them since about 1900. (It is a deep historical question why all this beautiful work of the generation of mathematicians active in 1870-1900 was not carried on in succeeding generations!) The 19-th century works on differential equations were dominated by analogies with algebraic equations. From this point of view, a Bäcklund transformation is analogous to what is called a correspondence in algebraic geometry. It is such a basic idea, that it must have extensive physical significance and application. Now, the traditional idea of "symmetry" of a differential equation--which has, of course, played such a strong role in recent work in elementary particle physics-- is a special case. One understands roughly how it carries over to quantum mechanics. How do Bäcklund transformations appear in quantum mechanics? No one knows. Worse, no one seems to have even thought about it. It is clear to me that understanding this point is the key to understanding the meaning of <u>quantum</u> solitons. (Classically, the "solitons" seem to be transforms under Bäcklund transformations of some sort of "ground state" wave.) I am very amused (and depressed!) that many (presumably) serious and competent physicists have proposed finding "quantum solitons" by means of "perturbation theory", often in models that do not even have <u>classical</u> soliton solutions. This is a good illustration of the mindlessness with which aggressive and competitive physicists jump into a subject with a fashionable label and--by the Gresham-Hermann Law--drive out serious and <u>long-term</u> creative mathematical thinking.

No one knows what Bäcklund transformations look like for differential equations involving a higher number of independent variables than two. Presumably, they are less likely to appear parameterized with discrete indices; of course, if they depended on continuous parameters, this could be even more interesting for elementary particles, since these continuous parameters could then perhaps be "quantized", à la Born-Sommerfeld quantization conditions.

Another interesting facet to the Bäcklund phenomenon is that only certain types of equations seem to admit Bäcklund transformations of a certain type. For example, the Sine-Gordon equation

$$\frac{\partial^2 \phi}{\partial t^2} - \frac{\partial^2 \phi}{\partial x^2} = \sin \phi \tag{1}$$

<u>does</u>, its "perturbation theoretic" approximation

$$\frac{\partial^2 \phi}{\partial t^2} - \frac{\partial^2 \phi}{\partial x^2} = \phi - \frac{\phi^3}{3!} \tag{2}$$

does not! Again, I believe that this sort of observation is enormously significant for elementary particle physics (and damning to the physicists' delusion of discovering anything worthwhile in this direction from the usual standard and mindless "perturbation theory" of a quantum field theory treatise.) It seems to suggest (to me) that, instead of elementary particles being in some sense a "generic" property-almost any differential equation with the right symmetry properties will do, according to present dogma--it is very <u>special</u>. This would not be surprising. Nature always seems to pick (at least for physics) rather exceptional and isolated

PREFACE

mathematical structures. This is perhaps even at the heart of the "unreasonable effectiveness" of mathematics in physics, that Wigner has noted. Notice that Einstein's gravitational field equations are not any old choice of equation with the right symmetry properties, but the ones that seem to be <u>intimately</u> and <u>uniquely</u> defined by association with a set of <u>geometric</u> ideas.

Another facet to the theory of non-linear Partial Differential Equations that is completely ignored in speculation about Fundamental Physics (or 99.99...% so) is its possibly intimate relation to <u>global</u> properties. The .001% exception is a brilliant suggestion by D. Finkelstein [1] that "fields" define mappings (at fixed time) of R^3 plus the point at infinity, i.e., the three sphere S^3, into certain types of spaces; these then should have <u>homotopy</u> invariants. In particular, the fields which <u>presumably define</u> elementary particles seem to involve mappings into Lie groups, even <u>compact</u> Lie groups. This fits in magnificently with E. Cartan's theorem that the <u>third</u> homotopy group of any simple compact Lie group is isomorphic to the <u>integers</u>. (The second homotopy group is trivial.) This suggests a new type of "quantum number" which he called a <u>kink</u>. When one begins to think whether this is a "quantum" or a "classical" phenomenon, one realizes how completely and shockingly inadequate has been the mathematical understanding of quantum mechanics.

Finkelstein's suggestion throws a brilliant light on the difference between equations (1) and (2). The first involves a mapping

$$\phi \to \sin \phi$$

implicating S^1 and $SO(2,R)$, while the second involves nothing, from this Finkelstein point of view. We thus get a small insight into how perturbation theory might be completely wrong from a global and geometric point of view. The physicists are so involved (in the 18-th century manner) with "formulas", which should be "expanded in terms of coupling constant", that the possibility that they are dealing with <u>mappings</u> between spaces with definite topological properties, which cannot be butchered in this brute-force way, does not seem to have entered their consciousness. I am especially annoyed about this, because some of my books (e.g., LAQM, VB) were meant to introduce this way of thinking to them; I have been told that they are "too mathematical". Of course, the mathematicians object to them (see the reviews of Chernoff and Marsden in the "Bulletin of the American Mathematical Society) precisely because I attempted to build some sort of a bridge between mathematical and physical ideas. (I suppose C & M would answer that I did not build a very good bridge. Of course, its all too easy to be high-minded and critical about the imperfection of attempts at doing this sort of applied work--it is the mathematicians vice. My criticism of C & M's review--for what it is worth--is that they did not seem to be very interested in the details of the physics, or at least expected me to hand it to them (and to the mathematical world in general) on a silver platter.)

To end this harangue, I hope the reader can see that I feel very strongly that many parts of physics will be in an impasse until certain difficult and underdeveloped mathematical material is understood in the right physical context. Despite the negative tone of all of this, I am not pessimistic--we are im possession of many of the tools, and they must be put to work. This involves getting back some of that old-time religion, with mathematicians and physicists talking to each other again.

Finally, I would like to thank Frank Estabrook and Hugo Wahlquist, of the Jet Propulsion Laboratory, for explaining their brilliant idea to me. Professors J. Corones of Iowa State University and Kent Harrison of Brigham Young University have also been of great help. I have learned almost all I know about Elementary Particles from Professor S. Glashow of Harvard, although we completely disagree. Karin Young has been of great help through her beautiful and efficient typing.

TABLE OF CONTENTS

Page

PREFACE iii

Chapter I: MAPPING ELEMENT SPACES AND SYSTEMS OF
PARTIAL DIFFERENTIAL EQUATIONS 1

1. Notation and Conventions 1
2. Notation for Partial Derivatives 6
3. Mapping Element (Jet) Spaces 9
4. Systems of Differential Equations and their
Solutions 14
5. Cauchy-Kowalewski Systems 20
6. A Geometric and Group-Theoretic Interpretation
of Cauchy-Kowalewski Systems 26

Chapter II: GENERAL SOLUTIONS OF DIFFERENTIAL EQUATIONS 31

1. Introduction 31
2. The General Definitions 31
3. The General Solution of Systems of Ordinary,
First Order, Differential Equations 38
4. The General Solution of Linear, Constant
Coefficient Partial Differential Equations.
Remarks on Quantum Field Theory 47
5. General Solutions of Foliations 55
6. General Solution of the Two-Dimensional Wave
Equation 60

Chapter III: MAYER SYSTEMS AND FOLIATIONS 65

1. Introduction 65
2. The Classical Notion of a Mayer Differential
System 66
3. Integrability of Mayer Systems 72
4. Remarks about Global Properties of Completely
Integrable Mayer Systems 77

Chapter IV: COMPLETE FAMILIES OF SOLUTIONS OF
DIFFERENTIAL SYSTEMS 81

1. Introduction 81
2. Complete Families of Solutions of First
Order Equations 83
3. Complete Families of Solution Manifolds of
Exterior Systems 87

CONTENTS

Page

Chapter V: SOME GENERAL IDEAS OF GALOIS THEORY   91

1. Introduction   91
2. The General Idea of Galois Theory   92
3. The Galois Theory of Field Extensions and Polynomials   101
4. The Classical Approach to Galois Theory   118
5. The Galois-Picard-Vessiot Theory of Ordinary, Linear Differential Equations   121
6. Solvability by Quadratures and Solvable Lie Groups   127
7. Reducibility of the Galois-Picard-Vessiot Group and Reducibility of Differential Operators   129
8. Principal Lie Systems   132
9. The Prolonged Group of a Lie Group   135
10. The Galois-Picard-Vessiot Groups of Principal Lie Systems   142

Chapter VI: LIE AND BÄCKLUND SYMMETRIES OF DIFFERENTIAL EQUATION SYSTEMS   147

1. Introduction   147
2. Mapping Element Spaces Defined in Terms of Linear Differential Operators   148
3. The Lie Differential Form System   153
4. Lie Symmetries and Bäcklund Symmetries of Differential Equation Systems   157

Chapter VII: THE PROLONGATION STRUCTURE OF ESTABROOK AND WAHLQUIST, THE DIFFERENTIAL GEOMETRY OF SOLITONS, AND CARTAN-EHRESMANN CONNECTIONS   165

1. Introduction   165
2. The Basic Notion of "Prlongation"   166
3. A Prolongation of the Korteweg-de Vries Equation   168
4. Systems which Admit a Pseudoconservation Law which is Quadratic in the Pseudopotential   171
5. Lie Algebra Valued Differential Forms   176
6. Lie Algebra Valued One-Forms and Cartan-Ehresmann Connections   183
7. The Bianchi Identity. The Frobenius Condition for Two-Forms   189

CONTENTS xi

Page

8. Quadratic Prolongations of Non-Linear Wave
   Equations in Two Independent Variables   191
9. The Geometric Foundation of the Inverse
   Scattering Technique   194
10. The Bäcklund Transformation in the Sense of
    Estabrook and Wahlquist   201

Chapter VIII: BÄCKLUND TRANSFORMATIONS   207

1. The Bäcklund Transformation of the Sine-Gordon
   Equation   207
2. Differential Equation Homomorphisms and
   Bäcklund Transformations   209
3. Linearization of Burger's Equation by Means
   of a Bäcklund Transformation   213

Chapter IX: THE LAPLACE-DARBOUX TRANSFORMATION, LINEAR
            BÄCKLUND TRANSFORMATIONS, AND THE INVERSE
            SCATTERING TECHNIQUE   217

1. Introduction   217
2. The Gauge Group and its Differential
   Invariants   218
3. The Laplace-Darboux Transform as a Linear
   Bäcklund Transformation   223
4. Some Physical Interpretations of Darboux' Work.
   The Klein and Sine-Gordon Equations   228

Chapter X: HIGHER DERIVATIVE CONSERVATION LAWS FOR
           SYMPLECTIC MANIFOLDS   239

1. Introduction   239
2. The Symplectic Structure on the Tangent
   Bundle to a Symplectic Manifold   241
3. The Prolongation Formula for Vector Fields   244
4. A General Prolongation Process for Differential
   Forms   251
5. A General Setting for the Theory of Conservation Laws   252

CONTENTS

Page

Chapter XI: BÄCKLUND TRANSFORMATIONS AS CONNECTIONS AND EXTERIOR DIFFERENTIAL SYSTEMS   255

1. Introduction   255
2. Bäcklund Transformations and Exterior Differential Systems   257
3. Introduction of a Connection   259
4. Bäcklund Transformations Determined by SL(2,R)-Connections   261
5. Bäcklund Transformations in Terms of Connections with Two-Dimensional Fibers   265
6. Another Two-Variable Connection Approach to the Sine-Gordon Bäcklund   267

Chapter XII: SOME BRIEF REMARKS CONCERNING THE ALGEBRAIC SETTING FOR THE THEORY OF EXTERIOR DIFFERENTIAL SYSTEMS   271

1. Introduction   271
2. Systems Generated Algebraically by Zero- and One-Forms   273
3. Systems Generated Algebraically by One-Forms and a Single Two-Form   274
4. Algebraic Study of Pairs of Skew Symmetric Bilinear Forms by Means of Kronecker's Theory of Pencils of Matrices   281
5. A General Invariant-Theoretic Setting   283
6. Algebraic Invariants of the Second Order Partial Differential Equation in Time Independent Variables   285

Chapter XIII: DEFORMATIONS OF EXTERIOR DIFFERENTIAL SYSTEMS AND SINGULAR PERTURBATION THEORY   289

1. Introduction   289
2. Singular Perturbation of Second Order, Linear, Constant Coefficient Ordinary Differential Equations   293
3. Singular Perturbation via the Ricatti Equation   295
4. Solution Subsets of Exterior Differential Systems   298

BIBLIOGRAPHY   305

FINAL REMARKS   307

Chapter I

MAPPING ELEMENT SPACES AND SYSTEMS OF
PARTIAL DIFFERENTIAL EQUATIONS

1. NOTATION AND CONVENTIONS

We continue to use the differential-calculus-on-manifolds notation and ideas that have been described in previous volumes. DGCV and GPS may be consulted for a systematic presentation. In this volume, I will attempt to adapt these notations so that they can be used in conjunction with the classical notation, e.g., as in Von Weber's encylopedia article.

In this volume, manifolds are denoted by letters like

$$X, Y, Z, \cdots .$$

Points of X are denoted by x, points of Y by y, etc.

Suppose X is a manifold of dimension n. Choose indices as follows, and the summation convention on these indices:

$$1 \leq i, j \leq n .$$

A <u>coordinate</u> system of real-valued functions for X will be denoted by

$$(x^i) .$$

Thus, $x^1, \ldots, x^n$ are real-valued, C^∞ functions on X, i.e., elements of $F(X)$. For example, if

$$X = R^n,$$

a point $x \in X$ is an ordered n-tuple

$$(a_1,\ldots,a_n)$$

of real numbers. The <u>Cartesian coordinate system</u> for R^n are the coordinates (x^i) such that:

$$x^i(a_1,\ldots,a_n) = a_i, \qquad 1 \leq i \leq n.$$

I will now introduce a notation that may be confusing, but which is necessary in order to keep close to the classical notation. If X is a manifold with coordinate system

$$(x^i),$$

the <u>point</u> x of X is denoted by

$$x = (x_1,\ldots,x_n) \in R^n.$$

Thus, the x_i are <u>really</u> real-valued functions on

$$X \times (\text{space of coordinate system for } X),$$

i.e.,

$$x_i(x, (x^{(\ )})) = x^i(x),$$

where $(x^{(\ )})$ denotes the coordinate system.

<u>Remark</u>. In this volume, I will attempt to systematically carry along "upper" and "lower" indices, i.e., "contravariant" and "covariant" indices, as in tensor analysis. The reason for this is that it is often very useful in complicated

calculations, and some parts of our subject are very heavily computational.

If X is a manifold, V(X) denotes the vector fields, i.e., the space of derivations of the algebra F(X). $F^r(X)$, for $r = 1,2,\ldots$, denotes the r-th degree differential forms.

$$\tau_s^r(X)$$

denotes the space of r-times contravariant, s-times covariant vector fields on X. In other words,

$$\tau_s^r(X) = \underbrace{V(X) \otimes \cdots \otimes V(X)}_{r \text{ times}} \otimes \underbrace{F^1(X) \otimes \cdots \otimes F^s(X)}_{s \text{ times}}$$

More generally, if

$$E \to X$$

is a fiber space, with X as base space, then

$$\Gamma(E)$$

denotes the space of cross-section maps ($\otimes$ denotes <u>tensor product of</u> F(X)-<u>modules</u>.)

The classical theory of differential equations usually deals with spaces of mappings. To keep as close as possible to these ideas, I will usually work in this context, although, of course, it would be little extra work to work with spaces of cross-sections, as was done in VB and GPS.

If X, Y are manifolds,

$\Gamma(X,Y)$ = space of maps

$\gamma: X \to Y$

It will be useful to keep as close as possible to the notation in Von Weber's article. (This is the reason I chose letters X and Z for manifolds, instead of $M, N, \cdots$ to which I am accustomed in other volumes.) To this end, suppose

(x^i) , $\qquad 1 \leq i, j \leq n = \dim X$

(z^a) , $\qquad 1 \leq a, b \leq m = \dim Z$

are coordinate systems for X and Z. As explained above, <u>points</u> x, z are denoted as follows:

$x = (x_1, \ldots, x_n)$

$z = (z_1, \ldots, z_m)$

A $\gamma \in \Gamma(X, Z)$ is then determined in these coordinates by real valued functions $f_a(x_1, \ldots, x_n)$ on R^n by the rule:

$$\gamma(x) = (f_1(x_1, \ldots, x_n), \ldots, f_m(x_1, \ldots, x_n)) \qquad (1.1)$$

Classically, this relation is usually denoted by:

$$z_a = f_a(x) , \qquad (1.2)$$

and we shall often use this convenient "abus de langage", as Bourbaki put it in his marvelously pedantic way.

A word is of course necessary about the precision with which we mean to deal with questions of domain of definition of maps, degrees of differentiability, etc. In modern differential geometry and topology, one usually works in the "category" of <u>globally defined</u>, C^∞ maps between C^∞ manifolds. This is the context in which I try to work, since it leads to the minimal number of notational monstrosities.

An alternative is to work with (real) analytic manifolds and <u>germs</u> of functions and maps, i.e., purely "local" concepts. Carried to its logical (and usually absurd) limits, in the context of the theory of "sheaves", this leads to grotesque algebraic abstractions and notational jargon, as in the work of Grothendieck and his disciples.

Of course, in the theory of differential equations these problems are especially acute because (except in certain restricted situations describable by ordinary differential equations) the relevant existence theorem (e.g., Cauchy-Kowalewski) only provides for "local", real-analytic solutions. The classical authors were purposely vague about this point, and I will follow their lead.

Thus, I will choose my <u>language</u> as if everything were globally defined and C^∞, leaving it to the reader's own skill in reading mathematics to sort things out.

Another confusing feature in the classical notation is the tendency to use <u>complex</u> variables and complex-analytical functions in certain situations, and usually in a rather abrupt and implicit way. From our modern "manifold" point of view, this is often not so routine. (Of course they usually thought in terms of "formulas", so that it is simple enough to substitute complex variables for real.)

2. NOTATION FOR PARTIAL DERIVATIVES

In any general treatment of the theory of partial differential equations along classical lines, a good deal of attention must be paid to notation <u>that is best suited to calculation</u>, since the non-trivial content often involves difficult and extensive computation. Here is a system I will use here, adapted from that used by Von Weber, and other classical authors.

Let X and Z be manifolds with coordinates (x^i), (z^a). Suppose a map

$$\gamma: X \to Z$$

MAPPING ELEMENTS

is determined by functions $f_a(x)$. As in the classical literature, this is denoted by:

$$z_a = f_a(x_1,\ldots,x_n) \tag{2.1}$$

We often use a <u>vector notation</u> for this:

$$z = f(x) \tag{2.2}$$

(Of course, (2.2) takes on a literal meaning as a map $R^n \to R^m$, if

$$z = (z_1,\ldots,z_m) \in R^m,$$

$$x = (x_1,\ldots,x_n) \in R^n \;.)$$

In the usual calculus notation, partial derivatives are denoted by:

$$\frac{\partial z_a}{\partial x^i} \equiv \frac{\partial f_a}{\partial x^i}$$

$$\frac{\partial^2 z_a}{\partial x^i \partial x^j} \equiv \frac{\partial^2 f_a}{\partial x^i \partial x^j},$$

and so forth.

Here is a notation which is better suited to extensive calculation. Let

$$\alpha = (\alpha_1,\ldots,\alpha_n) \in I_+ \times \cdots \times I_+ \equiv I_+^n$$

be a sequence of non-negative integers. (I denote the ring of integers, I_+ consists of the non-negative ones.)

Set:
$$|\alpha| = \alpha_1 + \cdots + \alpha_n \qquad (2.3)$$

Here is the basic notational abbreviation:

$$\partial_\alpha z_a = \frac{\partial^{|\alpha|} z_a}{\partial x_1^{\alpha_1}\ldots \partial x_n^{\alpha_n}} \qquad (2.4)$$

We also write this vectorially:

$$\partial_\alpha z \equiv \partial_\alpha f \qquad (2.5)$$

Here is another useful notation:

If $\alpha = (\alpha_1,\ldots,\alpha_n)$, set:

$$\alpha(i) = (\alpha_1,\ldots,\alpha_{i-1}, \alpha_i+1, \alpha_{i+1},\ldots,\alpha_n) \qquad (2.6)$$

Thus,

$$\frac{\partial}{\partial x^i}(\partial_\alpha z) = \partial_{\alpha(i)} z \qquad (2.7)$$

(Always with $z = f(x)$, of course.) Here is a differential form version of this relation that is very important:

$$d(\partial_\alpha z) = \partial_{\alpha(i)}(z)\, dx^i \qquad (2.8)$$

MAPPING ELEMENTS

3. MAPPING ELEMENT (JET) SPACES

The theory of jets, due in its modern form to C. Ehresmann, has proved to be a key idea linking classical ideas in differential geometry and differential equation theory. It is also very useful in physics and engineering, as I have tried to show in VB and GPS.

However, one petty, but substantial stumbling block to the theory becoming better known has been the name, "jet", which is misleading, particularly to a physicist or engineer.

Now, the concept occurs in implicit form in the work of Sophus Lie. (See the translations of some of Lie's work to be published in the Lie Theory series.) He calls them the space of "elements". Accordingly, I have replaced the name "jet" by the name "mapping element".

Let X, Z be manifolds with coordinates
$$(x^i), (z^a) .$$

Let $\gamma, \gamma': X \to Z$ be mappings, determined by the sets of functions:

$$\gamma: x \to z(x)$$

$$\gamma': x \to z'(x) .$$

Definition. γ and γ' meet to the r-th order at a point $x \in X$ if the following condition is satisfied:

$$\partial_\alpha z(x) = \partial_{\alpha'} z(x) \qquad (3.1)$$

for all $\alpha \in I_+^n$ such that $|\alpha| \leq r$.

Keep in mind what this means geometrically. For example, for

$$X = R = Z \quad.$$

Two maps γ, γ can be defined by their graphs

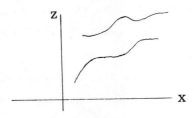

To say that (3.1) holds at x at $r = 0$ is to say that they meet at x:

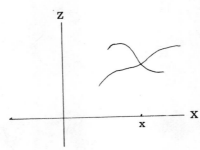

MAPPING ELEMENTS

To say that (3.1) holds for $r = 1$ is to say that they have the same tangent line

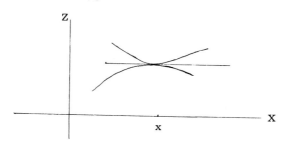

since, from elementary geometry the tangent line has the slope $(dz/dx)(x)$. One says in this case that they <u>meet at</u> x <u>to the first order</u>. Similarly, if (3.1) holds for $r = 2$, it means that <u>the curves meet to the second order at</u> x.

<u>Definition</u>. Consider the following equivalence relation on $X \times \Gamma(X,Z)$:

(x,γ) is equivalent to (x',γ')
if and only if
$$x = x' \qquad (3.2)$$
γ and γ' satisfy (3.1), i.e.,
meet to the r-th order at p.

The quotient of $X \times \Gamma(X,Z)$ by this equivalence relation is called the <u>r-th order mapping element</u> space, denoted by

$M^r(X,Z)$

(In the usual Ehresmann theory, $M^r(X,Z)$ is denoted by $J^r(X,Z)$, and is called the space of r-<u>th order jets</u>.)

To prove that $M^r(X,Z)$ is a manifold, we shall construct a coordinate system for it. For $\alpha \in I_+^n$ such that $|\alpha| \leq r$, let ∂_α be the maps

$$\partial_\alpha z^a : X \times \Gamma(X,Z) \to R \qquad (3.3)$$

defined by the following formula:

$$\partial_\alpha z^a(x,\gamma) = \partial_\alpha z_a(x) \qquad (3.4)$$

<u>Warning</u>. Remember that "z" plays a different role on both sides of (3.4). $\partial_\alpha z^a$ denotes the map. $x \to (z_a(x))$ are the functions defining the map γ.

Notice that the maps $\partial_\alpha z^a$ are <u>constant on the equivalence classes of the equivalence relation</u> (3.2), hence pass to the quotient to define maps, which we denote as follows:

$$\partial_\alpha z^a : M^r(X,Z) \to R \quad .$$

Similarly, let x^i, z^a be regarded as functions:

$$M^r(X,Z) \to R \quad .$$

(In words,

MAPPING ELEMENTS

$x^i(x,\gamma)$ = i-th component of x

$z^a(x,\gamma)$ = a-th component of $z(x) = (z_1(x),\ldots,z_m(x))$.)

Theorem 3.1. The functions $(x, z, \partial z)$ defined in this way define a <u>coordinate system</u> for $M^r(X,Z)$.

Exercise. Prove Theorem 3.1.

Exercise. Suppose new coordinates

$$(x'^i, z'^a)$$

are chosen for X and Z. Compute the functions $\partial_\alpha z'$, at least for small values of r.

The $(x^i, z^a, \partial_\alpha z)$, with contravariant indices represent real-valued functions on the mapping element spaces. We can also introduce the covariant objects

$$(x_i, z_a, \partial_\alpha z) \qquad (3.5)$$

to represent "points" of $M^r(X,Z)$. Thus, a point (3.5) (an element of a Euclidean space R^N, for a certain value of N which depends on n, m and r), represents the point $\partial^r \gamma(x) \in M^r(X,Z)$ such that:

$$\begin{aligned}
x^i(\partial^r \gamma(x)) &= x_i \\
z^a(\partial^r \gamma(x)) &= z_a \\
\partial_\alpha z^a(\partial^r \gamma(x)) &= \partial_\alpha z_a
\end{aligned} \qquad (3.6)$$

Alternate notation for the $\partial_\alpha z^a$ and $\partial_\alpha z_a$, used extensively in the classical literature, are:

$$z^a_\alpha \; ; \quad z_{a,\alpha} \; ; \quad p_{\alpha,a} \tag{3.7}$$

<u>Definition</u>. The coordinates $(x, z, \partial z)$ for $M^r(X, Z)$ defined in this way are called a <u>Lie coordinate system for the</u> mapping element space.

<u>Remark</u>. Again, the reader should keep in mind that I am referring to the Encyclopedia article by Von Weber for further references to the "classical literature".

4. SYSTEMS OF DIFFERENTIAL EQUATIONS AND THEIR SOLUTIONS

Let X, Z, $M^r(X,Z)$, $\Gamma(X,Z)$ be as above.

Here is a general definition.

<u>Definition</u>. Let N be a manifold. A subset $N' \subset N$ is said to be <u>defined by differential relations</u> if there are a finite number

$$f_1, f_2, \ldots$$

of (e.g., C^∞) differentiable real-valued functions on N such that:

$$N' = \{p \in N : 0 = f_1(p) = f_2(p) = 0 \cdots \} \tag{4.1}$$

MAPPING ELEMENTS 15

For short, we will say that such a subset is a <u>differentiable subset</u> of N.

<u>Remark</u>. Do not take what is meant by "differentiable" too precisely and/or literally. In the classical literature, the concept is appropriately fluid, sometimes meaning "analytic", in the technical sense, sometimes only a finite number of derivatives existing, sometimes analytic with certain "singularities", etc. Sometimes in differential topology one tries to speak of "smooth" objects, with precise categorical meaning. Perhaps this will ultimately be the way to do it, but at the moment I encourage the reader to follow the classical spirit and keep his mind open on this point. The way that the best 19-th century mathematicians worked was to be purposely vague, but in a subtle way. Hence this style is very appropriate.

Return to the case where N is the mapping element space
$$M^r(X,Z) \ .$$

<u>Definition</u>. A system of differential equations, with domain X and range Z, is a differentiable subset of
$$M^r(X,Z) \ .$$

Typically, such a system is denoted by S. Here is an alternate, and often used, classical terminology:

>The variables of X are the <u>independent variables</u> of S.
>The variables of Z are the <u>dependent variables</u> of S.

More concretely, in terms of the Lie coordinates

$$(x, z, \partial z)$$

for $M^r(X, Z)$, one can think of a differential equation system S as defined by a set of relations of the form:

$$\begin{aligned} F_1(x, z, \partial z) &= 0 \\ F_2(x, z, \partial z) &= 0 \\ &\vdots \end{aligned} \qquad (4.2)$$

The integer r is called the <u>order</u> of the system.

<u>Warning</u>. We shall later on define the concept of "strict order", which is closer to what order means in the classical literature.

Here is another basic notion.

<u>Definition</u>. Let $\gamma \in \Gamma(X, Z)$ be a map: $X \to Z$. For $x \in X$, let

$$\partial\gamma(x) \in M^r(X,Z)$$

be the equivalence class (defined by the equivalence relation (3.2)) to which

$$(x,\gamma) \in X \times \Gamma(X,Z)$$

belongs. As x varies, let

$$\partial\gamma: x \to \partial\gamma(x)$$

be the resulting map $X \to M^r(X,Z)$. It is called the r-th order derivative of γ.

Remark. It is assumed that r is fixed. If it is to vary, one can denote it by

$$\partial^r\gamma \quad .$$

Ehresmann calls this map the r-jet of γ, and denotes it by:

$$j^r(\gamma) \quad .$$

It is, of course, important to relate this abstract definition of "derivative" to the classical ideas. To do this, choose coordinates

$$(x^i), (z^a)$$

for X, Z, and represent γ in these coordinates by functions of the form:

$$z_a = f_a(x) \quad .$$

Associate with $\gamma \in \Gamma(X,Z)$, $x \in X$, the numbers:

$(x, z, \partial_\alpha z)$,

with $|\alpha| \leq r$.

We see that:

$\partial \gamma$ is the map which is represented, in the Lie coordinate system for $M^r(X,Z)$, by the functions

$x \to (x, z(x), \partial_\alpha z(x)): |\alpha| \leq r$.

<u>Definition</u>. Let $S \subset M^r(X,Z)$ be an r-th order differential system. A map $\gamma: X \to Z$ is a <u>solution</u> of S if the following condition is satisfied

$$\partial \gamma(X) \subset S \qquad (4.3)$$

<u>Remark</u>. If the system S is defined by relations of the form (4.2), then condition (4.3) takes the following more traditional form:

$$F_1(x, z(x), \partial z(x)) = 0$$
$$F_2(x, z(x), \partial z(x)) = 0 \qquad (4.4)$$
$$\vdots$$

MAPPING ELEMENTS

Remark. If the functions $F_1, F_2, \ldots$ are independent of the partial derivative variables, notice that the system reduces to a set of ordinary equations. For example, if the F are polynomials, the space of solutions can be described by the methods of algebraic geometry. A good deal of the motivation of the classical theory is to generalize concepts which are known for such ordinary equation systems to differential systems. For example, Lie intended much of his work on the group-theoretic properties of _differential_ equations to be a generalization of Galois' work on the group-theory of _algebraic_ equations. The subject now called "differential algebra" (see Kolchin [1]) in a sense represents a realization of this idea, but the _geometric_ aspects of the classical theory have been completely ignored.

One major difference between algebraic equations and differential equations is the relative weakness of the existence theorems for solutions of the latter kind. In dealing with algebraic equations the methods of either differential calculus (e.g., the Implicit Function Theorem) or algebraic geometry are quite effective in seeing roughly how the set of solutions is "parameterized". However, for differential equations the only general tool of this sort which is available is the Cauchy-Kowalewski theorem, which has several defects. Here are some:

    a) It only applies to <u>analytic</u> systems. In fact, one of the highlights of the modern theory of differential equations is H. Lewy's example of a C^∞ equation with <u>no solution at all</u>.

    b) It is too closely tied to choice of coordinates.

Cartan's theory of Exterior Differential Systems is, in a sense, the coordinate-free answer to b), but it is not an ideal tool when treating differential systems given in <u>their classical form</u>. It is often strikingly superior to the classical methods when dealing with problems arising from geometry which are presented directly in coordinate-free form. See Cartan's book, "Les systèmes exterieures..."

5. CAUCHY-KOWALEWSKI SYSTEMS

Section 22.2 of the Von Weber Encylopedia article presents an extensive and interesting review of the classical work on the <u>general</u> theory of existence of solutions of (real analytic) differential systems. The simplest general result of this theory is the <u>Cauchy-Kowalewski theorem</u>, which is also the only part which is well-known today. (For example, the work of Riquier [1]

MAPPING ELEMENTS   21

and Tresse [1] seems to merit reexamination in the
light of modern concepts.)

Definition. A differential system S involving independent
variables

$$x = (x_1, \ldots, x_n) ,$$

dependent variables

$$z = (z_1, \ldots, z_m) ,$$

is a Cauchy-Kowalewski system if it is of the following
form:

$$\partial_1^{r_1} z_1 = f_1(x, z, \partial z)$$
$$\vdots$$
$$\partial_1^{r_1} z_m = f_m(x, z, \partial z)$$

(5.1)

with the following conditions satisfied:

a) The functions $f_1, \ldots, f_m$ on the right hand
side of (5.1) are real-analytic.

b) They involve only the partial derivatives

$$\partial_{\alpha_1} z_1, \ldots, \partial_{\alpha_m} z_m$$

such that:

$$|\alpha_1| \leq r_1, \ldots, |\alpha_m| \leq r_m ,$$ (5.2)

and:

$$\alpha_1 \neq (r_1, 0, \ldots, 0)$$
$$\vdots \qquad (5.3)$$
$$\alpha_m \neq (r_m, 0, \ldots, 0)$$

Remark.

∂_1 denotes the partial derivative $\frac{\partial}{\partial x_1}$

∂_1^s denotes $\frac{\partial^s}{\partial x_1^s}$

$\alpha = (i_1, \ldots, i_n) \in I_+^n$, so that

$$\partial_\alpha = \frac{\partial^{|\alpha|}}{\partial x_1^{i_1} \ldots \partial x_n^{i_n}}$$

A key qualitative fact is that the <u>number of unknowns is equal to the number of equations</u>.

In words, conditions (5.2)-(5.3) say that the partial derivatives of the z_a occur on the right hand side of (5.1) only up to order r_a, and that the equations are "solved" for the derivatives on the left hand side of (5.1).

Theorem 5.1. (<u>Cauchy</u>, <u>Kowalewski and Darboux</u>). Suppose the functions on the right hand side of (5.1) are real analytic. Then, there is (locally) a unique real analytic solution

$z(x)$

of (5.1), such that the "Cauchy data" functions

$z(0,\vec{x})$,

$$\frac{\partial z_1}{\partial x_1}(0,\vec{x}), \ldots, \frac{\partial z_1}{\partial x_1^{r_1-1}}(0,\vec{x})$$

$$\vdots \qquad\qquad\qquad (5.4)$$

$$\frac{\partial z_m}{\partial x_1}(0,\vec{x}), \ldots, \frac{\partial z_m}{\partial x_1^{r_m-1}}(0,\vec{x})$$

are arbitrary functions which are given in advance. ($\vec{x}$ denotes $(x_2,\ldots,x_n) \in R^{n-1}$.)

Remark. For the complete proof, see Courant-Hilbert [1] or Goursat [1].

The idea is to first find a formal power series solution, then to prove it converges by the <u>method of majorants</u>. The formal power series part is quite easy. Prescribing the Cauchy data functions (5.4) fixes the values of <u>all</u> the derivatives with respect to $\vec{x}$ at the point $0 \in R^n$. It also fixes the values at 0 of

$$\partial_1^{i_1} z_1, \ldots, \partial_1^{i_m} z_m$$

for:
$$1 \leq i_1 \leq r_1-1, \ldots, 1 \leq i_m \leq r_m-1 \ .$$

But now the differential equations (5.1) themselves determine the values at 0 of the remaining derivatives

$$\partial_1^i a$$

It is also easy to see that the resulting formal power series satisfies the differential equation <u>if it converges</u>.

To prove convergence, one uses the <u>method of majorants</u>, devised by Cauchy. Suppose given two functions, or two formal power series,

$$g_1(x), \ g_2(x)$$

on R^n. Let us say that

g_2 <u>majorizes</u> g_1, denoted by

$$g_1 \ll g_2 \ ,$$

if the following condition is satisfied:

$$|\partial_\alpha g_1(0)| \leq \partial_\alpha g_2(0) \qquad (5.5)$$
$$\text{for all} \quad \alpha \in I_+^n \ .$$

Using the hypothesis that the data are real analytic (and an estimate proved by Cauchy for the coefficients of a convergent power series), the right hand side of (5.1) is replaced by functions f_1', f_2' which majorize f_1, f_2 and which have the property that the resulting Cauchy-Kowalewski system can be solved explicitly. Similarly, the Cauchy data are majorized. One then shows that the solution of the system obtained by replacing the data by their majorizations majorizes the formal power series solution of the original system, hence the formal power series solution of (5.1) converges.

This argument--which I think of as one of the most ingenious in mathematics--is the first example of the technique of a priori estimates, that now plays a dominant role in the theory of partial differential equations. (See Hörmander [1].) The idea is to find inequalities satisfied by broad classes of solutions of a differential equation without detailed knowledge of the solution. Another typical example is the maximum principle for the Laplace equation--a harmonic function in a region is always less in absolute value than its boundary values. Often a key feature in an existence or approximation theorem for a differential equation involves such estimates. They have an obvious practical importance, e.g., for numerical calculation on computers. Unfortunately, there is usually no smooth qualitative technique for finding such estimates (as the one we have just described for Cauchy-Kowalewski systems), but only grubby calculations. (This is probably

as it should be--modern abstract mathematics offers an
<u>approach</u> to the key and difficult problems but, as in
mountain climbing, the smooth approach must at some point
be abandoned for real work.)

<u>Exercise</u>. Show that, by adding certain derivatives of the
dependent variables as new unknowns, a Cauchy-Kowalewski
system can be converted to a first order system.

6. A GEOMETRIC AND GROUP-THEORETIC INTERPRETATION OF CAUCHY-KOWALEWSKI SYSTEMS

In this section, we shall deal with first order
Cauchy-Kowalewski systems. Ordinary, differential
equations form one special type. For example, suppose
there is but one independent variable, which we call "t",
for obvious physical reasons. (For simplicity, we suppose
that the right hand sides of the systems considered in
this section do <u>not</u> depend explicitly on t.) Thus, the
system takes the following form:

$$\frac{dz}{dt} = f(z) \tag{6.1}$$

The Lie theory of ordinary differential equations (see
DGCV and Vessiot [1]) provides a convenient and group-
theoretic interpretation of (6.1). Introduce the vector
field

MAPPING ELEMENTS

$$\underset{\sim}{v} = f^a \frac{\partial}{\partial z^a} \qquad (6.2)$$

on R^m. The solutions of (6.1) are, geometrically, the <u>integral curves</u> of $\underset{\sim}{v}$, and, group-theoretically, the <u>orbits</u> of the one-parameter group of diffeomorphisms of R^m generated by the vector field $\underset{\sim}{v}$.

Now, we shall change notation slightly from those used in Section 5 to write a first order Cauchy-Kowalewski system as follows:

$$\frac{\partial z}{\partial t} = f(z, \vec{x}, \partial_x z) \qquad (6.3)$$

Here, $z = (z_1, \ldots, z_m)$ are the dependent variables. The space of the independent variables are written as a Cartesian product:

$$x = (t, \vec{x}) \; ,$$

$$t \in R, \quad \vec{x} = (x_2, \ldots, x_n) \; . \qquad (6.4)$$

<u>Remark</u>. Physically, this suggests that X is "space-time", with t the <u>time</u> and $\vec{x}$ the <u>space</u> variables. In fact, a good deal of the notation for the pure mathematical theory of systems of this type comes from physics. In classical physics, the problems are mainly from continuum and fluid mechanics, but there are now a whole new

class of such problems provided by quantum field theory. These "neo-classical" problems provide a major challenge to contemporary mathematics.

Let $\vec{X}$ denote the space of the variables $\vec{x}$, i.e., R^{n-1}. Recall that

$$\Gamma(\vec{X}, Z)$$

denotes the space of all maps

$$\vec{X} \to Z \ .$$

Let us suppose that the Cauchy problem may be solved globally, i.e., given a map

$$\gamma: \vec{x} \to z(\vec{x}) \ \epsilon \ \Gamma(X, Z) \ ,$$

there is a unique solution of (6.3), reducing to $z(\vec{x})$ at $t = 0$. Denote by

$$\gamma_t$$

the map

$$\vec{x} \to z(t, \vec{x})$$

of $\vec{X} \to Z$. We see then that the Cauchy-Kowalewski system (6.3), and the initial map γ, determine a curve

$$t \to \gamma_t$$

in $\Gamma(\vec{X}, Z)$.

MAPPING ELEMENTS

> This curve may be interpreted as the orbit of a one-parameter group of transformations on $\Gamma(\vec{X}, Z)$.

The expression on the right hand side of (6.3) may be interpreted as the <u>infinitesimal generator of this group</u>. We shall now discuss its geometric nature.

Write (6.3) as:

$$\frac{\partial z^a}{\partial t} = f^a(z(t,\vec{x}), \vec{x}, \partial_{\vec{x}} z(t,\vec{x})) \tag{6.5}$$

Define a map

$$\underset{\sim}{v}: M^1(\vec{X}, Z) \to T(Z) \tag{6.6}$$

as follows:

$$\underset{\sim}{v}(\vec{x}, z, \partial z) = f^a(\vec{x}, z, \partial z) \frac{\partial}{\partial z^a} \tag{6.7}$$

<u>Remark</u>. Recall that $M^1(X,Z)$ denotes the <u>first order mapping elements</u> of maps $: X \to Z$. $T(Z)$ denotes the <u>tangent bundle to</u> Z. We see that (6.7) defines a map

$$\underset{\sim}{v}: M^1(\vec{X}, Z) \to T(Z) \quad.$$

$\underset{\sim}{v}$ reduces to the concept of vector field when the space $\vec{X}$ is <u>zero dimensional</u>. See my paper, "E. Cartan's geometric theory of differential equations" for more detail about the geometry of these "generalized vector fields", and their use in Cartan's theory of exterior differential systems.

Chapter II

GENERAL SOLUTIONS OF DIFFERENTIAL EQUATIONS

1. INTRODUCTION

The classical concept of "general solution" of a differential equation is notoriously difficult to understand and/or make precise in terms of modern mathematics. In this chapter I shall <u>briefly</u> try to explain what is meant (roughly following Von Weber), and then proceed to study various related special topics.

2. THE GENERAL DEFINITIONS

Let X and Z be manifolds. For each integer r, let

$$M^r(X,Z)$$

denote the mapping element space, with domain X, co-domain Z.

$M(X,Z)$ denotes the space of maps

$$X \to Z \ .$$

A typical element of $M(X,Z)$ is denoted by

$$\underline{z} \ .$$

$\partial^r \underline{z}$ denotes its r-th order prolongation, a map

: $X \to M^r(X,Z)$

Let
$$DE \subset M^r(X,Z)$$
be a differentiable subset of $M^r(X,Z)$. It defines a <u>differential equation</u> (with X domain, Z codomain). A $\underline{z} \in M(X,Z)$ is a <u>solution</u> of DE if:
$$\partial^r \underline{z}(X) \subset DE \ .$$
$\underline{S}$ denotes the space of all solutions.

Suppose that the subset DE is defined by setting the functions
$$f_1,\ldots,f_s : M^r(X,Z) \to R \ .$$
equal to zero. A $\underline{z} \in M(X,Z)$ then belongs to $\underline{S}$ if and only if:
$$\begin{aligned} f_1(\partial^r \underline{z}(x)) &= 0 \\ &\vdots \\ f_s(\partial^r \underline{z}(x)) &= 0 \end{aligned} \quad (1.1)$$
$$\text{for all } x \in X \ .$$

We can now differentiate both sides of the relation (1.1). We see that $\partial^{r+1}\underline{z}$ satisfies a set of equations. This defines a differential equation system
$$(DE)^1 \subset M^{r+1}(X,Z) \ ,$$

GENERAL SOLUTION

called the <u>first prolongation of</u> DE. This process can be iterated, defining <u>second</u>,... prolongations

$$(DE)^2 \subset M^{r+2}(X,Z)$$
$$\vdots$$

<u>Definition</u>. A differential equation system

$$(DE)' \subset M^{r+p}(X,Z)$$

is called a <u>subsystem</u> of DE if

$$(DE)^p \subset (DE)' \tag{1.2}$$

<u>Remark</u>. Here is what this means. Consider DE defined by Equation (1.1). Differentiate (1.1) p times, to obtain the system $(DE)^p$. (1.2) now means that the equations defining $(DE)'$ are chosen as a subset of those defining $(DE)^p$.

Now, we arrive at the definition of "general solution":

<u>Definition</u>. Let GS be a subset of $M(X,Z)$ such that:

$$GS \subset \underline{S} \quad , \tag{1.3}$$

i.e., each $\underline{z} \in GS$ is a solution of (1.1). GS is said to define a <u>general solution</u> of DE if the following condition is satisfied:

If (DE)' is another system of differential equations such that each $z \in GS$ is a solution of (DE)', then (DE)' is a sub-system of DE, in the sense defined above. (1.4)

<u>Remark</u>. This definition is ascribed by Von Weber, to Ampère. Here is its more classical formulation. Suppose that GS is a set of maps : $X \to Z$, parameterized by arbitrary functions and parameters. GS is the "general solution" of (1.1) if, after differentiating and eliminating the arbitrary functions and parameters, one obtains the relations (1.1) and their derivatives, but <u>no other relations</u>.

In the classical literature, there is an alternative definition, due to Darboux. Namely, a subset $GS \subset M(X,Z)$ is a <u>Darboux general solution</u> if each solution of the "Cauchy problem" of (1.1) can be obtained by appropriate "specialization" of the arbitrary constants or functions in GS.

There are certain difficulties in understanding this concept in full generality. What is meant by the "Cauchy problem"? By "arbitrary constants or functions, and their specialization"? Von Weber asserts that a "general solution" in Darboux' sense is also one in Ampère's sense, but not necessarily conversely.

GENERAL SOLUTION

Here is an idea which might make some of this precise. Let Y,W be manifolds, V a finite dimensional vector space. A "general solution" of (1.1) might be defined as a map

$$M(Y,V) \times W \to \underline{S}$$

whose image fills up "most" of S, with perhaps some "singular" pieces left out. This map also should satisfy certain "smoothness" conditions. The dimension of V equals what the classical authors call the "number of arbitrary functions", and the dimension of W is the "number of arbitrary constants".

Let us now turn from these extremely general concepts to more concrete situations.

2. GENERAL SOLUTIONS OF FIRST ORDER SCALAR DIFFERENTIAL EQUATIONS

Suppose that:

$$X = Z = R \quad.$$

Denote the variable on X by x, the variable on Z by z. Suppose $\underline{S}$ consists of the maps $\underline{z}: X \to Z$ which satisfy:

$$\frac{dz}{dx} = f(z,x) \quad. \tag{2.1}$$

Let Λ be another copy of R, with variable λ, and let

$$\phi : X \times \Lambda \to Z$$

be a map, denoted by:

$$\phi(x,\lambda) = z(x,\lambda) .$$

For each $\lambda \in \Lambda$, let

$$\underline{z}_\lambda \in M(X,Z)$$

be the map $x \to z(x,\lambda)$. Let

$$GS = \{\underline{z}_\lambda : \lambda \in \Lambda\} \qquad (2.2)$$

When does (2.2) define a General Solution of (2.1)? The first, obvious condition is that:

$$\frac{\partial z}{\partial x} = f(x,z) \qquad (2.3)$$

This says that, for each λ, $x \to z(x,\lambda) \equiv \underline{z}_\lambda(x)$ is a solution of (2.1).

Suppose also that:

$$\frac{\partial z}{\partial \lambda} \neq 0 \qquad (2.4)$$

This is a suitable "non-singularity" condition. It guarantees that (locally) z can be replaced by λ as a dependent variable. In this way, we can reduce everything to the case:

GENERAL SOLUTION

37

$$z = \lambda .$$

In this case, we have:

$$f(x,z) = 0 ,$$

i.e., the differential equation (2.1) takes the form:

$$\frac{dz}{dx} = 0 . \tag{2.5}$$

The prolonged systems are then:

$$\frac{d^2z}{dx^2} = 0$$
$$\vdots \tag{2.6}$$

Now suppose that (DE)' is a system of differential equations such that $\{\underline{z}_\lambda : \lambda \in R\}$ is a set of solutions of (DE)'. Suppose one of the equations in (DE)' is:

$$h\left(x, z, \frac{dz}{dx}, \ldots\right) = 0 .$$

Substitute in the relation:

$$z(x) = \lambda .$$

We have:

$$h(x,\lambda,0,\cdots) = 0$$

Then, $h(x,z',z'',\cdots)$ can be written as a polynomial in the $z',z'',\cdots$ with <u>coefficients that are functions of</u> x,z. But, this is <u>precisely</u> what is meant by saying that (DE)' is a subsystem of (2.5)-(2.6)!

3. THE GENERAL SOLUTION OF SYSTEMS OF ORDINARY, FIRST ORDER, DIFFERENTIAL EQUATIONS

The simple-minded material in Section 2 can now readily be generalized to more general ordinary differential equations. Here is one such setting.

Let X be a one-dimensional manifold, and let Z be an n-dimensional one. Choose indices, and the summation convention as follows:

$$1 \leq i,j \leq n \; .$$

Let x denote a coordinate on X, and let (z^i) be coordinates on Z.

Suppose given a first order differential equation, applied to maps $X \to Z$, of the following form:

$$\frac{dz^i}{dx} = f^i(x,z) \tag{3.1}$$

On $X \times Z$, introduce the following Pfaffian forms:

$$\omega^i = dz^i - f^i \, dx \tag{3.2}$$

The solutions of (3.1) may then be considered as the one-dimensional submanifolds of $X \times Z$ on which the forms (3.2) restrict to zero, and which are transversal to x.

Introduce another n-dimensional manifold Λ, with coordinates

$$(\lambda^i) \; .$$

GENERAL SOLUTION

Consider a map

$$\phi: X \times \Lambda \to X \times Z$$

of the form:

$$\phi(x,\lambda) = (x, z(x,\lambda)) \quad . \tag{3.3}$$

<u>Definition</u>. Such a map ϕ defines a <u>general solution</u> for the differential equation (3.1) if both of the following conditions are satisfied:

$$\phi^*(\omega^i) \text{ is a linear combination of the } d\lambda^i. \tag{3.4}$$

$$\phi \text{ is a local diffeomorphism} \tag{3.5}$$

Let us see what these conditions mean. Using (3.4),

$$\phi^*(\omega^i) = d(z^i(x,\lambda)) - f^i(x, z(x,\lambda)) \, dx$$

$$= \frac{\partial z^i}{\partial x} dx + \frac{\partial z^i}{\partial \lambda^j} - f^i \, dx \quad .$$

Hence, (3.4) implies that:

$$\frac{\partial z^i}{\partial x} = f^i(x, z(x,\lambda)) \quad . \tag{3.6}$$

(3.6) means that, for fixed λ, the map $x \to z(x,\lambda)$ $\equiv \underline{z}_\lambda(x)$ is a solution of (3.1). (3.6) means that

$$\det\left(\frac{\partial z^i}{\partial \lambda^j}\right) \neq 0$$

i.e., the relations $\lambda \to z$ can be inverted to express λ in terms of z.

(3.4) can be most readily interpreted in terms of the theory of <u>Pfaffian systems</u>. Let P be the Pfaffian system on $X \times Z$ generated by the ω^i. Let P' be the Pfaffian system on $X \times \Lambda$ generated by the $d\lambda^i$. Condition (3.4) means that:

$$\phi^*(P) = P',$$

ϕ is an <u>isomorphism</u> (or <u>equivalence</u>, in the sense of Lie and Cartan) between P and P'. In particular, ϕ carries a solution submanifold of P' into a solution submanifold of P. Alternately, ϕ carries a solution of

$$\frac{d\lambda^i}{dx} = 0$$

into a solution of (3.1). The functions

$$\phi^{-1}*(\lambda^i)$$

are <u>conserved</u> along the solutions of (3.1). (The classical terminology, e.g., in Vessiot [1], is that they are <u>integrals of the system</u>.)

GENERAL SOLUTION

41

Remark. This definition of "general solution" fits in with Lie's ideas of groups of symmetries of differential equations.

Let

$$\phi: X \times \Lambda \to X \times Z$$

be a map of form (3.3) which satisfies condition (3.4). (We do not necessarily assume that ϕ is a local diffeomorphism. In this way, we can--as explained later--cover some of the classical material concerning "singular solutions".) Let G be a group, which acts as a group of diffeomorphisms of $X \times \Lambda$, such that:

$$g^*(P') = P'$$

for all $g \in G$.

(In other words, g is of the form

$$g: (x',\lambda') \to (x(x',\lambda'), \lambda(\lambda'))$$.

Suppose that G can be made into a group of diffeomorphisms of $X \times Z$, so that ϕ intertwines the two actions. (Of course, if ϕ is a diffeomorphism, this can be done in precisely one way. If ϕ is a local diffeomorphism, G can be made to act <u>locally</u> on $X \times Z$.) This action of G in $X \times Z$ is then a <u>group of symmetries</u> of the differential equation (3.1), in the sense that

$$g*(P) = P$$

for all $g \in G$.

Here is a typical case.

Example 1. <u>The differential equation whose general solution is the set of all circles in the plane</u>.

This example--taken from Kowalewski [1], page 2--is a typical one in Lie theory. Start off with an independent variable x, a dependent variable y. (x,y) together parameterize R^2.

For each $(\lambda^1, \lambda^2, \lambda^3) \in R^3$, consider the circle in R^2:

$$(x-\lambda^1)^2 + (y-\lambda^2)^2 = (\lambda^3)^2 \tag{3.7}$$

with center (λ^1, λ^2), and radius λ^3. It is readily seen that the functions $y(x;\lambda)$ resulting from solving (3.7) for y as a function of x satisfy the following third order differential equation:

$$\left(1 + \left(\frac{dy}{dx}\right)^2\right) \frac{d^3y}{dx^3} - 3 \frac{dy}{dx} \left(\frac{d^2y}{dx^2}\right)^2 = 0 \tag{3.8}$$

Set:

$$z^1 = y, \quad z^2 = \frac{dy}{dx}, \quad z^3 = \frac{d^2y}{dx^2}$$

$$z = (z^1, z^2, z^3).$$

GENERAL SOLUTION

Then, (3.8) is equivalent to a differential equation system of the form (3.1). (3.7) clearly defines a "general solution". This means that a map

$$(x,\lambda) \to (x, z \equiv z(\lambda,x))$$

can be found by solving (3.7) for y as a function of x, and differentiating twice with respect to x.

Now, let

$$G = SO(1,3) ,$$

the Lorentz group. It acts (via conformal transformations) on R^2, in such a way as to take circles into circles. (Really, this is only a "local" action, but it can be readily handled globally by compactifying R^2 to make the sphere.)

Remark. This action is nothing but the group of <u>linear fractional transformations</u> or <u>Mobius transformations</u> familiar from complex function theory. To define it, identify $(x,y) \in R^2$ with the complex number $w = x + iy$. Identify an element $g \in G$ with a 2×2 complex matrix

$$\begin{pmatrix} a & b \\ c & d \end{pmatrix}$$

of the determinant one. Then,

$$g(x,y) = \frac{aw + b}{cw + d} .$$

This geometric action determines a <u>transitive</u> action on Λ, the parameter space of the circles.

<u>Exercise</u>. Determine the subgroup H of $G = SO(1,3)$ such that
$$\Lambda = G/H .$$

<u>Remark</u>. In order to make literal sense of this, Λ itself must be <u>compactified</u> by adding in the <u>straight lines</u> of R^2, as limiting cases of circles.

<u>Exercise</u>. Let G act in the natural geometric way (through conformal transformations) in $R^2 \equiv$ space of variables (x,y). Let G act via Lie prolongation on $X \times \Lambda$. Show, using general arguments and <u>also</u> explicit calculation, that G acts on a group of symmetries on the differential equation (3.8).

<u>Exercise</u>. Using general principles (and the above Remark), explain why the straight lines are also solutions of (3.8).

Example 2. <u>The Clairault equation</u>.

It is the example traditionally used to explain the relation between the "singular" and "general" solution.

Consider the space L of all straight lines in R^2. It is a two-dimensional manifold. It can be parameterized as follows:

GENERAL SOLUTION

An element of L, with parameters (a,b), has the equation:

$$z = ax + b .$$

Consider a one-dimensional submanifold of L, defined by giving b as a function f(a) of a. Restricting to this submanifold, we obtain a one-parameter family

$$\{z = ax+f(a): a \in R\} \tag{3.9}$$

of straight lines in R^2.

We can now find a first order differential equation satisfied by each of the straight lines in the set (3.9). To do this, follow the classical procedure: Differentiate 3.9 with respect to the variable x, and eliminate a:

$$\frac{dz}{dx} = a , \quad \text{i.e.,}$$

$$z = x \frac{dz}{dx} + f\left(\frac{dz}{dx}\right) \tag{3.10}$$

This is a <u>Clairault differential equation</u>. (3.9) defines its <u>general solution</u>.

<u>Problem</u>. The group $G = SL(3,R)$ acts (locally) on R^2 as <u>linear fractional transformations</u> or <u>the projective group</u>. It maps lines into lines, hence acts geometrically on L. Determine which one-dimensional subspaces of L of the

form (3.9) are orbits of one-parameter subgroups of G. Determine the corresponding differential equations, and verify that they admit symmetry groups as one-parameter groups of projective transformations. Is every first order differential equation in R^2 which admits a one-parameter projective group as symmetries of this form?

Remark. With these two examples we encounter a very interesting feature of the classical theory which has hardly been worked on at all in the modern literature. Namely, one can consider differential equations as <u>defined</u> and <u>classified</u> by giving certain natural families of geometric objects which form their general solution. Naturally, the most interesting such geometric objects are those which are acted on by a Lie group. (This is why straight lines and circles are important!) Thus, the Clairault equation is very "natural" from this point of view!

Another interesting feature is the close tie at this point with algebraic geometry, which also studies continuous families of geometric objects of this type.

4. THE GENERAL SOLUTION OF LINEAR, CONSTANT COEFFICIENT PARTIAL DIFFERENTIAL EQUATIONS. REMARKS ON QUANTUM FIELD THEORY

Now we turn to another situation where the "general solution" can be discussed systematically. It is, of course, of great importance in mathematical physics.

Let X, Z, W be finite dimensional vector spaces. Let

$$(x^i) \, , \quad 1 \leq i, j \leq n = \dim X$$

be a basis for X^d. For

$$\alpha \in I_+^n \, ,$$

i.e., $\alpha = (\alpha_1, \ldots, \alpha_n)$, with $\alpha_1, \ldots, \alpha_n$ non-negative integers, set:

$$\partial\alpha \;=\; \frac{\partial^{|\alpha|}}{\partial x_1^{\alpha_1} \ldots \partial x_n^{\alpha_n}}$$

Consider a <u>linear</u>, constant coefficient differential operator

$$D: M(X, Z) \to M(X, W)$$

of the following form:

$$D(\underline{z}) \;=\; \sum_{\alpha \in I_+^n} A^\alpha \partial_\alpha \underline{z} \tag{4.1}$$

GENERAL SOLUTION

In (4.1), $\underline{z}$ is an element of $M(X,Z)$, i.e., a map $X \to Z$. For each α,

A^α is a linear map $Z \to W$.

(A^α is zero for all but a finite number of α's.)

Let $y \in X^d$, $z \in Z$. Set:

$$\underline{z}(x) = z e^{y(x)} \qquad (4.2)$$

Now,

$$\partial_\alpha (e^{y(x)}) = \alpha(y) e^{y(x)} , \qquad (4.3)$$

where

$$y \equiv y_1 x^1 + \cdots + y_n x^n , \qquad (4.4)$$

$$\alpha(y) \equiv y_1^{\alpha_1} \cdots y_n^{\alpha_n} \qquad (4.5)$$

(Thus, $y \to \alpha(y)$ is a polynomial map : $X^d \to R$. See Vol. VIII of IM.)

Hence, with $\underline{z}$ defined by (4.2) (which is, physically, called a **plane wave**),

$$D(\underline{z}) = \sum_\alpha A^\alpha(z) \alpha(y) e^{y(x)} \qquad (4.6)$$

In particular,

$D(\underline{z}) = 0$ if and only if:

$$\sum_\alpha A^\alpha(z) \alpha(y) = 0 \qquad (4.7)$$

GENERAL SOLUTION

Let E be the subset of

$$(y,z) \in X^d \times Z$$

which satisfy condition (4.7). Map

$$E \to X^d$$

as follows:

$$(y,z) \to y \quad .$$

Call it the <u>projection map</u>, and denote it by π. In words, π is the Cartesian product projection map

$$X^d \times Z \to X^d \quad ,$$

restricted to the subset E.

For fixed $y \in X^d$, the fiber

$$\pi^{-1}(y)$$

may be identified with a <u>vector space</u>, namely with the $z \in Z$ which satisfies (4.7). E is therefore a <u>vector bundle</u> over X^d.

We need a way to "parameterize" the fibers of E. In general, this is very hard to do precisely. (See Ehrenpreis [1].) We <u>assume</u> it can be done in the following way. Let

$$Y \subset X^d$$

be a submanifold such that, there is a vector space V and a map

$$\phi: Y \times V \to E$$

such that, for each $y \in Y$,

$$\phi(y,V) = \pi^{-1}(y) \quad .$$

Set:

$$\phi(y,v) = (y,\gamma_y(v))$$

For fixed y, γ_y is a map $V \to Z$.

Thus, for

$$(y,v) \in Y \times V \quad ,$$

we can construct a solution

$$\underline{z}_{y,v}$$

of the differential equation

$$D = 0$$

using the formula:

$$\underline{z}_{y,v}(x) = \gamma_y(v) e^{y(x)} \qquad (4.8)$$

We can now use the <u>linearity</u> of D and the resulting <u>principle of superposition</u> for solutions of $D = 0$. To do this, choose a volume element differential form

$$dy$$

on Y, and a map

$$\underline{v}: Y \to V$$

GENERAL SOLUTION

Set:

$$\underline{z}(\underline{v}) = \int_Y \underline{z}_{y,\underline{v}(y)} \, dy \tag{4.9}$$

$\underline{v} \to \underline{z}(\underline{v})$ is then a linear map:

$$M(Y,V) \to (\text{solutions of } D = 0) \subset M(X,V) \tag{4.10}$$

If Y is chosen <u>appropriately</u>, this may be considered as the <u>general solution</u> of $D = 0$.

This is only a sketch. See Ehrenpreis [1] for more detail. In practice, there are many difficulties. For example, the integral on the right hand side of (4.9) may not converge, and one has to develop a <u>generalized function</u> (or <u>distribution</u>) formalism.

These ideas are very important for <u>quantum field theory</u>. (See LAQM, VB, and Vol. VI of IM.) Let

$$\underline{S} \subset M(Z,W)$$

be the space of <u>solutions</u> of the differential equation

$$D = 0 \ .$$

$\underline{S}$ is to be regarded as the set of <u>states</u> of the <u>classical field</u> defined by the differential operator D. <u>Observables</u> are then real-valued functions on $\underline{S}$.

As we have seen, $\underline{S}$ may be identified--with the exception of certain "singular" pieces--with

$$M(Y,V) \ .$$

The observables may then be identified with sets of functions on M(Y,V).

Remark. For example, a "system with a finite number of degrees of freedom" may be considered as the case where D is an <u>ordinary</u> differential <u>operator</u>, i.e.,

dim X = 1 .

In this case, Y is <u>zero dimensional</u>, hence, M(Y,V) is a discrete direct sum of a finite number of copies of V. In this case, the <u>observables</u> become a set of functions on a finite dimensional manifold.

Unfortunately, what it means to <u>quantize</u> such a classical field is not a well-defined process. Here is one interpretation, which I suggested in LAQM. Pick out some subset of the observables, and impose some sort of a algebraic structure on the observables--for example, a Lie algebra structure analogous to <u>Poisson bracket</u>. Then attempt to represent some suitable Lie subalgebra of this Poisson bracket algebra <u>irreducibly</u> by skew-Hermitian operators on a Hilbert space. The elements of this Hilbert space are called the <u>quantum states</u>. Unfortunately, there is nothing particularly unique or even well-defined about this process, especially for <u>systems of an infinite number of degrees of freedom</u>.

GENERAL SOLUTION

I can, however, briefly indicate the process that is used in practice. We have seen that $\underline{S}$ is identified, via the "general solution", with a space

$$M(Y,V)$$

of mapping of a manifold Y into a vector space V. Let

$$M(Y,V^d)$$

be the space of maps of Y into the dual space of V. Each

$$\gamma^d \in M(Y,V^d)$$

defines a <u>linear form</u> on $M(Y,V)$:

$$\gamma \to \langle \gamma^d, \gamma \rangle = \int_Y \gamma^d(y)(\gamma(y))\, dy \quad .$$

In particular, the elements of γ^d become <u>classical observables</u>, i.e., linear maps on the classical states. Often, V^d carries a skew-symmetric bilinear form

$$\omega: V^d \times V^d \to R \quad .$$

This form can be used to define a skew-symmetric bilinear form on

$$M(Y,V^d):$$

$$\underline{\omega}(\gamma_1^d, \gamma_2^d) = \int_Y \omega(\gamma_1^d(y), \gamma_2^d(y))\, dy \qquad (4.11)$$

<u>Remark</u>. Typically, (4.11) is equivalent to what physicists call the <u>canonical commutation relations</u> for fields.

One can now try to represent the Heisenberg Lie
algebra formed by these linear classical observables.
(Even this is non-trivial mathematically, since in this
infinite dimensional setting the irreducible representations of the Heisenberg algebra are not unique. In practice, one proceeds by choosing a complex vector space
structure on V^d, which defines an annihilation-creation
operator formalism, and a unique Fock space representation.)

How to deal with the quantization of non-linear
differential operators D is even murkier. (In practice,
physicists work with an elaborate perturbation theory
formalism to utilize what they know about the linear case,
but it should be clear to anyone with mathematical intuition involving non-linear differential equations that this
can be expected to be satisfactory in only very restricted
and accidental circumstances. Indeed, elementary particle
physics--which is in a sense "applied quantum field theory"--
has been, at a fundamental level, at an impasse for twenty
five years for precisely this reason. The renormalization
techniques that worked in the quantum electrodynamics case
(which is the "accidental" situation referred to above)
do not generalize correctly. Presumably, this is because
the underlying equations involve "truly" non-linear phenomena
and there are no more cheap conquests here.

GENERAL SOLUTION

Unfortunately, physicists are "pragmatists", and not willing to seriously undertake new mathematical exploration. (For example, the recent work on non-linear wave propagation, (Whitham [1]), particularly the "soliton" concept, is very interesting as a clue to what a "truly" non-linear quantum field elementary particle theory should look like.)

Even for non-linear differential equations, the "general solution" should be a useful concept for quantization. In the next few sections, I shall follow this up for ordinary differential equations. A paper by E. Cartan [1] on the Calculus of Variations (which seems to be almost completely unknown and mysterious) gives certain interesting clues. To develop his ideas involves several differential-geometric concepts, which we briefly explore.

5. GENERAL SOLUTIONS OF FOLIATIONS

Let X be a manifold. Let

$$V(X)$$

denote the vector fields on X. It forms a module over $F(X)$ (the real valued functions on X) and a real Lie algebra, with the Jacobi bracket as Lie algebra operation.

A <u>vector field system</u> is defined as an $F(X)$-submodule

$$V \subset V(X) \quad .$$

For $x \in X$, set:

$$V(x) = \{A(x): A \in V(X)\}$$

$$\equiv \text{ set of values of } V \text{ at } x.$$

V is said to be <u>non-singular if</u>:

    dim $V(x)$ is constant as x ranges over X.

V is said to be <u>completely integrable</u> if:

$$[V,V] \subset V .$$

Given such a completely integrable, non-singular vector field system V, the global version of the <u>Frobenius complete integrability theorem</u> (see DGCV) provides, through each point $x \in X$, a <u>solution submanifold</u> $Y(x)$ of V which is <u>maximal</u>. It is called the <u>leaf</u> of V through x. This decomposition of X into leaves is said to define a <u>foliation</u> of X.

In particular, we will blur the distinction between V and the leaf decomposition it determines, and call the whole business a <u>foliation</u>, also taking the "non-singularity" for granted.

Definition. A <u>general solution</u> for the foliation V is a pair

$$(Y, \Lambda)$$

GENERAL SOLUTION   57

of manifolds, and a diffeomorphism

$$\phi: Y \times \Lambda \to X$$

such that:

> For each $\lambda \in \Lambda$, the submanifold
> $\phi(\lambda, Y)$ is a leaf of V. (5.1)

In other words, a "general solution" is an isomorphism between the given foliation and a Cartesian product foliation. The Frobenius complete integrability theorem for foliation (see DGCV) asserts that they exist <u>locally</u>. Of course, they may not exist globally.

Here is the relation to more classical ideas. Let T and Z be manifolds, with coordinates

$$(t^i), \ (z^a)$$
$$1 \leq i,j \leq n = \dim T$$
$$1 \leq a,b \leq m = \dim Z \ .$$

Consider a system of partial differential equations of the form:

$$\frac{\partial z^a}{\partial t^i} = f_i^a(t,z) \tag{5.2}$$

(As we shall see later on, in the 19-th century this was called a <u>Mayer system</u>.)

To convert these equations into a vector field system, set:

$$X = T \times Z .$$

$$\omega^i = dz^a - f_i^a dt^i \qquad (5.3)$$

$$V = \{A \in V(X): \omega^i(A) = 0\} . \qquad (5.4)$$

The integrability conditions

$$[V,V] \subset V$$

in terms of vector field systems are equivalent to the classical integrability systems for the system (5.2), obtained by differentiating both sides:

$$\frac{\partial f_i^a}{\partial t^j} + \frac{\partial f_i^a}{\partial z^b} f_j^b = \frac{\partial f_j^a}{\partial t^i} + \frac{\partial f_j^a}{\partial z^a} f_i^b \qquad (5.5)$$

If conditions (5.5) are satisfied, then there is a solution

$$t \to z(t)$$

of (5.2), a map $T \to Z$, which is uniquely determined by taking on a value $z_0 \in Z$ at a point $t_0 \in T$.

The graph of this solution

$$t \to (t, z(t))$$

is then a submanifold of $T \times Z$. It is a leaf of the foliation V.

GENERAL SOLUTION

A general solution then is determined by a map

$$\phi: \Lambda \times Y \to T \times Z \quad.$$

of the form

$$\phi(\lambda, y) = (t(\lambda, y), z(\lambda, y))$$

Now, on each leaf the functions

$$t^i$$

are functionally independent. This means that, for fixed λ, the functions

$$y \to t^i(\lambda, y)$$

are functionally independent. Thus, t^i can be introduced (locally) as variables in place of y. ϕ is then determined by maps

$$\lambda \to z^i(t, \lambda) \quad,$$

which, for fixed λ, is a solution of (5.2). Thus, the general definition given above is one appropriate formulation of the "general" solution for the Mayer type of partial differential equation system.

<u>Remark</u>. As a bonus, we see a geometric interpretation of the parameters λ in the "general solution" notion: They parameterize the <u>leaves of the foliation</u> V.

6. GENERAL SOLUTION OF THE TWO-DIMENSIONAL WAVE EQUATION

The wave equation,

$$\frac{\partial^2 z}{\partial x_1^2} - \frac{\partial^2 z}{\partial x_2^2} = 0 \quad , \tag{6.1}$$

is, of course, one of the few partial differential equations which occurs in nature that can be solved explicitly. This is done by the following substitution

$$\begin{aligned} x &= x_1 - x_2 \\ y &= x_1 + x_2 \end{aligned} \tag{6.2}$$

Hence:

$$\frac{\partial}{\partial x_1} = \frac{\partial x}{\partial x_1} \frac{\partial}{\partial x} + \frac{\partial y}{\partial x_1} \frac{\partial}{\partial y}$$

$$= \frac{\partial}{\partial x} + \frac{\partial}{\partial y}$$

$$\frac{\partial}{\partial x_2} = -\frac{\partial}{\partial x} + \frac{\partial}{\partial y}$$

$$\frac{\partial^2}{\partial x_1^2} - \frac{\partial^2}{\partial x_2^2} = \left(\frac{\partial}{\partial x_1} + \frac{\partial}{\partial x_2}\right) \frac{\partial}{\partial x_1} - \frac{\partial}{\partial x_2}$$

$$= 4 \frac{\partial^2}{\partial x \partial y}$$

GENERAL SOLUTION

Hence,
$$z = f_1(x_1-x_2) + f_2(x_1+x_2)$$
is, for each arbitrary function $f_1(\ )$, $f_2(\ )$ of one variable, a solution of (6.1). We shall show that it is a general solution, in each of the senses (i.e., "Ampère" and Darboux") described earlier. (See Von Weber [1] and Forsyth [1] for the classical discussion.)

Let us first do the "Ampère" property. To this end, use the change of variables (6.2) to convert the problem to the following form:

$$Z = R = \text{space of variables } z$$
$$X = R^2 = \text{space of variables } (x,y).$$

$M^2(X,Z)$ is the second-order mapping element space, with Lie coordinates labelled:

$$(x,\ y,\ z_{10},\ z_{01},\ z_{11},\ z_{20},\ z_{02},\ \ldots)$$

Recall what these definitions mean:

$$z_{10} = \frac{\partial z}{\partial x}, \quad z_{01} = \frac{\partial z}{\partial y}, \quad z_{11} = \frac{\partial^2 z}{\partial x \partial y},$$

and so forth. The differential equation is defined by the subset

$$DE \subset M^2(X,Z)$$

defined by the following conditions:

$$z_{11} = 0 \qquad (6.3)$$

For each pair $f_1(\ )$, $f_2(\ )$ of functions of one-variable, let

$$\underline{z}(x,y) = f_1(x) + f_2(y) \qquad (6.4)$$

Let

$$F(x,\ y,\ z_{m,n}\colon m,n = 0,1,\ldots) \qquad (6.5)$$

be a differential equation satisfied by each map $\underline{z}$ defined by (6.4), for <u>arbitrary</u> choice of f_1, f_2. Substituting (6.4) into (6.5) gives:

$$F(x,\ y,\ f_1(x)+f_2(y),\ f_1'(x),\ f_2'(y),\ f_1''(x),\ \ldots) = 0 \qquad (6.6)$$

Suppose that the function F in (6.5) involves r-th order derivatives. We must show that it vanishes on the subset of $M^r(X,Z)$ defined by the relations:

$$0 = z_{m,n} \qquad (6.7)$$

for: $m + n = r$, m and $n > 0$.

(This defines the <u>prolongation</u> of the system (6.3).)

Now, if

$$\underline{z}(x,y) = f_1(x) + f_2(y)\quad,$$

then

GENERAL SOLUTION

$$\frac{\partial^{m+n} z}{\partial x^m \partial y^n} = \begin{cases} 0 & \text{if } m \text{ and } n > 0 \\ f_1^{(m)}(x) & \text{if } m = 0 \\ f_2^n(y) & \text{if } n = 0 \end{cases}$$

By choosing f_1, f_2 as polynomials of degree r, it is obvious that we can make $f_1^{(m)}(x)$, $f_2^{(n)}(y)$ take arbitrary values. Hence, F vanishing on

$$\partial^r \underline{z}(X)$$

for arbitrary f_1, f_2 implies that

F vanishes on the subset (6.7) of $M^r(X, Z)$,

which completes the proof that

$$z = f_1(x_1 + x_2) + f_2(x_1 - x_2)$$

is the general solution in the Ampère sense.

Exercise. Discuss the general solution property in the Darboux sense, i.e., show that the Cauchy problem for non-characteristic Cauchy data can be solved by particular choice of f_1, f_2.

Remark. Of course, this example is atypical in the sense that every solution is part of the general solution. For non-linear equations, this would not be typical--in fact, such phenomena might be very important in physics, related to "particle-like" behavior to be associated with wave equations.

Chapter III
MAYER SYSTEMS AND FOLIATIONS

1. INTRODUCTION

It is well-known that certain types of partial differential equations may be solved by means of successions of ordinary differential equations. The simplest general systems of this type are classically called Mayer systems. They are more-or-less equivalent to the geometric structures called foliations in the contemporary literature. They play a prominent role in the theory of Lie groups. In fact, the first modern treatment of the theory of foliations--in Chevalley's book [1946]--was stimulated by the need to develop an adequate foundation for the theory of Lie groups.

Since Chevalley's book, the theory of foliations has mainly been developed by topologists, as a generalization of the global theory of ordinary differential equations. There remain many problems of a mixed geometric-topological, local-global, nature that are related to problems of differential geometry, mathematical physics, differential equation theory, control and systems theory.

This chapter deals with a more restricted matter of the relations between foliation theory and certain classical ideas in the general theory of partial differential equations, but the reader should keep in mind that a wider context is in the background.

2. THE CLASSICAL NOTION OF A MAYER DIFFERENTIAL SYSTEM

Let X, Z be an in the previous chapter, i.e., X is a manifold, Z is a real vector space. Let

(x^i) be coordinates for X

(z^a) be linear coordinates for Z.

Let $M^r(X,Z)$ be the space of r-th order mapping elements. Let

$$(x^i, z^a, \partial_\alpha z^a)$$

be the Lie coordinates for $M^r(X,Z)$.

$$\alpha = (\alpha_1, \ldots, \alpha_n) \in I_+^n$$

is a sequence of non-negative integers, with:

$$|\alpha| \equiv \alpha_1 + \cdots + \alpha_n \leq r .$$

<u>Definition</u>. A differential equation system $S \subset M^r(X,Z)$ is a <u>Mayer system</u> if it is defined, in these Lie coordinates, by equations of the following form:

$$\partial_\alpha z^a = f_\alpha^a(x, z, \partial_\beta z) \tag{2.1}$$

for each $\alpha \in I_+^n$ such that $|\alpha| = r$, with derivatives ∂_β on the right hand side of order $\leq (r-1)$. In other words, the functions on the right hand side of (2.1) depend on the variables

$$(x, z, \partial_\beta z): |\beta| \leq r-1 \quad . \qquad (2.2)$$

In words, a <u>Mayer system of order</u> r is a system in which <u>all</u> derivatives of order r are given as functions of lower order derivatives.

Here is a coordinate-free way of defining this concept. There is a natural projection map

$$\pi^r: M^r(X,Z) \to M^{r-1}(X,Z) \quad , \qquad (2.3)$$

namely, one which "forgets" the derivatives of order r. (In other words, if $\gamma, \gamma' \in \Gamma(X,Z)$ meet to order r at x, they meet to order r-1, hence there is a projection map between the equivalence classes, which is just the map (2.3).) We see immediately from the definition that, in the Lie coordinates

$$(x, z, \partial_\alpha z) \quad , \quad |\alpha| \leq r \quad ,$$

for $M^r(X,Z)$, the

$$(x, z, \partial_\beta z) \quad , \quad |\beta| \leq r-1 \quad ,$$

are the pull-back under the map π^r of Lie coordinates on $M^{r-1}(X,Z)$. In other words, in these coordinates, the map π^r can be written as follows:

$$\pi^r(x,z,\partial_\alpha z: |\alpha|\leq r) = (x,z,\partial_\beta z: |\beta|\leq r-1) .$$

Thus, if S is the subset of $M^r(X,Z)$ defined by formulas (2.1), we see that π^r maps S in a <u>one-one</u> way <u>onto</u>

$$M^{r-1}(X,Z)$$

Hence, the inverse of (π^r restricted to S) is a map

$$\phi: M^{r-1}(X,Z) \to M^r(X,Z)$$

such that

$$\pi^r \phi = \text{identity} .$$

Such a map is, of course, called a <u>cross-section</u> of the fiber space (2.3). Hence, we have proved the following

<u>Theorem 2.1</u>. An n-th order Mayer system is determined as a cross-section of the fiber space (2.3).

Notice one advantage of this alternate form of the definition: it is completely coordinate free and geometric.

Here is one basic geometric property of Mayer.

<u>Theorem 2.2</u>. Suppose that S is a given Mayer system, and that

$$\gamma, \gamma': X \to Z$$

are two maps which are both solutions of S. Suppose also

that X is a connected manifold, and that there is one point $x^0 \in X$ such that:

$$\partial_\beta \gamma(x^0) = \partial_\beta \gamma'(x^0) \qquad (2.4)$$

for all $|\beta| \leq r-1$.

Then,

$$\gamma = \gamma'.$$

Proof. The set of all points x of X such that

$$\gamma(x) = \gamma'(x)$$

is obviously a closed subset of X. (Recall that we are assuming that all maps are sufficiently "smooth"--in particular, that they are continuous, which forces the set of points where two are equal to be a closed subset.) Thus, to prove $\gamma = \gamma'$, we must prove that the set of points where they are equal is also an open subset of X. It will then follow from connectedness of X that the set is all of X, i.e., $\gamma = \gamma'$.

To do this we may obviously use a Lie coordinate system, that we label as follows:

$$(x, z, \partial_\beta z: |\beta| \leq r-1;\ \partial_\alpha z: |\alpha| = r).$$

Suppose that γ and γ' are determined in these coordinates by functions which we label

$$z(x),\ z'(x).$$

Let
$$t \to x(t)$$
be a curve in X, such that
$$x(0) = x^0 .$$
Then,
$$\frac{d}{dt} z(x(t)) = \partial_i z(x(t)) \frac{dx^i}{dt}$$

$$\frac{d}{dt} (\partial_i z)(x(t)) = \partial_{ij} z(x(t)) \frac{dx^j}{dt}$$
$$\vdots \qquad\qquad\qquad\qquad\qquad\qquad (2.5)$$

$$\frac{d}{dt} (\partial_\beta z)(x(t)) = (\partial_k \partial_\beta z)(x(t)) \frac{dx^k}{dt}$$

For $|\beta| = r-1$, we have, using the equations (2.1) defining the Mayer system,

$$\frac{d}{dt} (\partial_\beta z)(x(t)) = f_{(\partial_k \beta)}(z(x(t)), x(t), \partial_\beta(x(t))) \frac{dx^k}{dt}$$
$$(2.6)$$

Consider $x(t)$, hence dx/dt, as <u>given</u> functions of t. We may then look on (2.5) and (2.6) as a set of ordinary differential equations, to be solved for the functions

MAYER SYSTEMS

$$t \to (\partial_\beta z)(x(t)), \qquad |\beta| \leq r-1 \quad . \tag{2.7}$$

Notice that the functions

$$t \to (\partial_\beta z'(x(t)), \qquad |\beta| \leq r-1 \tag{2.8}$$

satisfy precisely the same set of ordinary differential equations as do the functions (2.7), since the $z(x)$, $z'(x)$ satisfy the same Mayer system (2.1). By (2.4), the functions of t, (2.7) and (2.8), have the initial values. By uniqueness of the solution of ordinary differential equations, they coincide for all t. In particular,

$$\begin{aligned} z(x(t)) &= z'(x(t)) \\ \text{for all} \quad t. \end{aligned} \tag{2.9}$$

Now, the curve $t \to x(t)$ was chosen arbitrarily. Hence, (2.5) implies:

$$z(x) = z'(x)$$
$$\text{for } \underline{\text{all}} \quad x \quad ,$$

which finishes the proof that

$$\gamma = \gamma' \quad .$$

Remark. Let $\Gamma(S)$ denote the set of all solutions of the Mayer system (2.1). We can "parameterize" $\Gamma(S)$ as follows:

Pick a point $x^0 \in X$ arbitrarily. Assign to $\gamma \in \Gamma(S)$ its $(r-1)$-st mapping element at x^0:

$$\partial^{r-1}\gamma(x^0) \in M^{r-1}(X,Z) \tag{2.10}$$

This defines a map

$$\Gamma(S) \to M^{r-1}(X,Z) \quad .$$

By Theorem 2.1 it is <u>one-one</u>. (The condition that it be <u>onto</u> requires integrability, which we deal with in the next section.) In the classical language, this fact can be expressed by saying that:

<u>The set of solutions of a Mayer system depends on a finite number of parameters</u>.

<u>Exercise</u>. Make precise the "number of parameters" involved.

3. INTEGRABILITY OF MAYER SYSTEMS

Keep the notation of Section 2. For

$$\beta = (\beta_1, \ldots, \beta_n) \in I_+^n$$

and i an integer between 1 and n, let

$$\beta(i) = (\beta_1, \ldots, \beta_{i-1}, \beta_i + 1, \beta_{i+1}, \ldots, \beta_n) \tag{3.1}$$

For

$$|\beta| < r-1 \quad ,$$

set:

$$\theta_\beta^a = dz_\beta^a - z_{\beta(i)}^a \, dx^i \tag{3.2}$$

MAYER SYSTEMS

Recall that the

$$\{\theta_\beta^a: |\beta| \leq r-1\}$$

define the Lie differential form system, which we denote by L.

Here is a standard concept from the theory of differential form systems. (See Volume IX of IM, for example.)

<u>Definition</u>. Let M be a manifold, P an F(M)-module of one-forms on M, which defines a differential form system. The system P is said to be <u>completely integrable</u> if:

$$dP \subset F^{(\ )}(M) \wedge P \qquad (3.3)$$

i.e., if the <u>Grassman ideal</u> forms generated by P is also a differential ideal.

Return to the case where M is the mapping element space

$$M^r(X,Z) \quad ,$$

and P is L, the Lie system of one-forms.

<u>Definition</u>. A differential equation system

$$S \subset M^r(X,Z)$$

is said to be <u>completely integrable</u> if the Lie differential form system L is completely integrable when restricted to S.

Warning. This is not what "complete integrability" means in the classical literature, e.g., Von Weber's Encyclopedia article. I have followed more closely to Cartan's terminology. He says that a system is "in involution" where the 19-th century authors say that it is "completely integrable".

Let us see what this condition means for a Mayer system. Using (3.2), we have:

$$d\theta^a_\beta = dx^i \wedge dz^a_{\beta(i)} \tag{3.4}$$

Case 1. $|\beta| < r-1$.

In this case, $\beta(i) \leq r-1$, hence we can write:

$$dz^a_{\beta(i)} = d\theta^a_{\beta(i)} + z^a_{\beta(i)(j)} dx^j \tag{3.5}$$

Substituting this into (3.4), we have:

$$d\theta^a_\beta = dx^i \wedge z^a_{\beta(i)(j)} dx^j + \cdots$$

(The terms $\cdots$ denote terms in the Grassman ideal of forms generated by the Lie form L.)

Now,

$$\beta(i)(j) = \beta(j)(i) \tag{3.6}$$

Hence, for $|\beta| < r-1$,

$$d\theta^a_\beta \equiv 0 \mod L . \tag{3.7}$$

(Of course, by this terminology, we mean that $d\theta^a_\beta$ belongs to the Grassman algebra ideal generated by L. This is standard algebraic terminology.)

So far, we have not used the differential equations. This is involved in:

Case 2. $|\beta| = r-1$.

Then,
$$|\beta(i)| = r \, .$$

Hence, using the relations (2.1) defining a Mayer system, we have, on S,

$$z^a_{\beta(i)} = f^a_{\beta(i)} \, , \qquad (3.8)$$

Combine (3.8) with (3.4). Restricted to S,

$$d\theta^a_\beta = dx^i \wedge df^a_{\beta(i)} \, . \qquad (3.9)$$

Now,

$$df^a_{\beta(i)} = \partial_j f^a_{\beta(i)} \, dx^i + \sum_{|\alpha| \leq r-1} \frac{\partial (f^a_{\beta(i)})}{\partial z^b_\alpha} \, dz^b_\alpha$$

$$(3.10)$$

$$= \partial_j (f^a_{\beta(i)}) \, dx^i + \sum_{|\alpha| \leq r-1} \frac{\partial (f^a_{\beta(i)})}{\partial z^b_\alpha} \, z^b_{\alpha(j)} \, dx^j + \cdots$$

Set:

$$T^a_{ij} = \partial_j(f^a_{\beta(i)}) + \sum_{|\alpha| \leq r-1} \frac{\partial f^a_{\beta(i)}}{\partial z^b_\alpha} z^b_{\alpha(j)} \qquad (3.11)$$

Definition. (T^a_{ij}) is called the <u>integrability tensor</u> of the Mayer system.

Theorem 3.1. The Mayer system (2.1) is <u>completely integrable</u> if and only if:

$$T^a_{ij} = T^a_{ji}, \qquad (3.12)$$

i.e., the integrability tensor is symmetric.

The proof follows from (3.11) and (3.9).

Exercise. Investigate the "tensorial" properties of T^a_{ij}, e.g., under changes in coordinates for X and Z. What about changes of coordinates which mix up coordinates of X and Z, i.e., changes of coordinates for the manifold structure of X × Z?

Exercise. Suppose (3.12) is satisfied, i.e., the Lie system L restricted to S is completely integrable. Show that the maximal solution submanifolds of L (as provided by the Frobenious Complete Integrability Theorem -- see DGCV, Chapter 8) locally are defined by maps X → Z which are solutions of the system S.

4. REMARKS ABOUT GLOBAL PROPERTIES OF COMPLETELY INTEGRABLE MAYER SYSTEMS

In general, if X and Z are manifolds, and if

$$S \subset M^r(X,Z)$$

is a system of partial differential equations, it is (in our present state of mathematical knowledge) hopeless to try to find reasonable and useful conditions that there exist globally maps

$$\gamma: X \to Z$$

which are solutions of X. There are many problems, e.g., in physics and differential geometry, where such information would be very valuable, but it is just not available. However, if S is a completely integrable Mayer system, the question of the existence of global solutions is not so inaccessible as in the general case. For example, one reason that the theory of Lie groups now has solid global foundations is that the differential equations one encounters there are of Mayer type!

Let L be the Lie differential form system on $M^r(X,Z)$. Restrict it to S. Recall that the condition that S be "integrable" is that L restricted to S be "completely integrable", i.e., define a foliation of S. The leaves of this foliation are then the mapping elements of maps

$$\gamma: \text{(open subsets of } X) \to Z$$

which are solutions of S.

Recall also that S is a submanifold of $M^r(X,Z)$ which is the image, under a cross-section map

$$M^{r-1}(X,Z) \to M^r(X,Z) \quad .$$

In particular, S is <u>transversal</u> to the fibers of the fiber space map

$$M^r(X,Z) \to M^{r-1}(X,Z) \quad .$$

Now, the <u>global version</u> of the Frobenious complete integrability theorem (see DGCV, Chapter 63) applies to L restricted to S. Let N be a <u>leaf</u> of the foliation of S this determines, i.e., N is a submanifold of S which is a maximal integral submanifold of L.

<u>Exercise.</u> Consider the prolongation map

$$M^r(X,Z) \to X \quad .$$

Show that this map, restricted to N is a <u>local diffeomorphism</u>. Show that it is a <u>diffeomorphism</u> if and only if there is a map

$$\gamma: X \to Z$$

which is a solution of S, such that N is the subset

$$\partial^r \gamma(X) \subset S \quad .$$

Remark. Here is the import. If one is able to show that
the projection map

$$N \to X$$

is a covering map, and that it is onto, then standard
techniques provide solutions

$$\gamma: X \to Z$$

of S, or perhaps generalizations of solutions such as
maps

$$\gamma': X' \to Z$$

of X' = the simply connected covering space of X which
locally looks like the solution of S. See DGCV, Chapter 21
for useful material concerning the analytical conditions
that the projection $N \to X$ (which is automatically a local
diffeomorphism) be a covering map. If curves in X starting at one point can be lifted to curve in N, then it is
a covering map, and everything is well-behaved. Such lift-
ings of curves are, as we have seen in Section 2 in connection with uniqueness, determined by ordinary differential
equations. Thus, what we need are conditions that these
ordinary differential equations have global solutions. One
such condition is known: If the equations are linear.

Exercise. Suppose X is a simply connected manifold, that
Z is a finite dimensional real vector space, and that

$$S \subset M^r(X,Z)$$

is a <u>linear</u> completely integrable Mayer system. Show that, for each point

$$(x,z) \in X,Z \quad ,$$

there is a map

$$\gamma: X \to Z$$

which is a solution of S, such that

$$\gamma(x) = z \quad .$$

Chapter IV

COMPLETE FAMILIES OF SOLUTIONS OF
DIFFERENTIAL SYSTEMS

1. INTRODUCTION

The topic covered by the title of this chapter is one of the concepts of classical differential equation theory that is now forgotten. (See Von Weber [1].) I now intend to describe it in modern language.

Let us recapitulate notation. X and Z are manifolds, called the <u>independent and dependent variable manifolds</u>.

$M(X,Z)$ denotes the space of maps

$$\underline{z}: X \to Z \quad .$$

For each positive integer r, let

$$M^r(X,Z)$$

denote the space of r-th order mapping elements. If

$$\underline{z} \in M(X,Z) \quad ,$$

$$\partial^r \underline{z}: X \to M^r(X,Z)$$

denotes its prolongation.

<u>Definition</u>. An (r-th order) <u>differential equation system</u> is defined as a subset

$$DE \subset M^r(X,Z) \quad ,$$

which is a <u>differentiable subset</u>, i.e., defined by setting a finite number of C^∞ functions equal to zero.

Suppose such a differential equation, denoted by DE, is given. A <u>solution</u> is a $\underline{z} \in M(X,Z)$ such that

$$\partial^r \underline{z}(X) \subset DE \quad .$$

Let $\underline{S}$ denote the space of such solutions.

Now, we come to the main concept.

<u>Definition</u>. A <u>complete family of solutions</u> of a DE is defined by a manifold Λ, and a <u>submersion</u> mapping

$$\pi: DE \to \Lambda$$

such that, for each $\lambda \in \Lambda$, there is a solution map $\underline{z}_\lambda \in \underline{S}$ such that:

$$\partial^r \underline{z}_\lambda(X) = \pi^{-1}(\lambda)$$

(In words, it is defined by a <u>fibration</u>--or, perhaps, more generally, a <u>foliation</u>--of the subset of the mapping element space, whose fibers are prolongations of solutions of the differential equation.)

We shall now study the main classical example.

COMPLETE FAMILIES 83

2. COMPLETE FAMILIES OF SOLUTIONS OF FIRST ORDER EQUATIONS

Now, suppose:

$$\dim X = 1, \quad \dim Z = 1,$$
$$r = 1, \quad \dim (DE) = 2n \ .$$

Let (x^i) be a coordinate system of X. Let (x^i, z, z_j) be the corresponding Lie coordinate system for $M^1(X,Z)$. Thus, DE is defined by a relation

$$f(x, z, z_j) = 0 \qquad (2.1)$$

between these variables.

Following the general idea described in Section 1, a <u>complete family of solutions</u> of DE is defined by giving the following data:

a) A manifold Λ
b) A submersion mapping

$$\pi: DE \to \Lambda \ .$$

c) Given, for each $\lambda \in \Lambda$, a $\underline{z}_\lambda \in M(X,Z)$ such that:

$$\pi^{-1}(\lambda) = \partial^1 \underline{z}_\lambda (X) \ .$$

Remark. Since

$$\dim(DE) = \dim M^1(X,Z) - 1$$

i.e., we are dealing with a <u>single</u> first order partial differential equation, and since:

$$\dim M^1(X,Z) = 2n+1 \;,$$

$$\dim (\text{fibers of } \pi = \dim X = n) \;,$$

we see that:

$$\dim \Lambda = n \;.$$

Let us see what this means in terms of coordinates. Let

$$(\lambda^i)$$

be a set of coordinates for Λ. Then, $\underline{z}_\lambda$ is determined by functions

$$(x,\lambda) \to z(x,\lambda) = z$$

of $X \times \Lambda \to R$.

They satisfy the following equation:

$$f\left(x,\, z,\, \frac{\partial z}{\partial x}\right) = 0 \qquad (2.2)$$

We must take into account the condition that π is a submersion mapping, i.e., that

$$\pi_*(T(DE)) = T(\Lambda) \;.$$

COMPLETE FAMILIES   85

Exercise. Show that this is equivalent to the condition:

$$\det\left(\frac{\partial^2 z}{\partial x^i \partial \lambda^j}\right) \neq 0 \ . \tag{2.3}$$

These ideas are most frequently encountered in the context of the Hamilton-Jacobi equation of mathematical physics. This corresponds to the case where the function of defining the DE is independent of z.
Relabel

$$z_i \text{ as } p_i \ ,$$

the "momenta". f is then a function

$$f(x,p)$$

of these "dual" or "conjugate" variables. In physics it is usually the Hamiltonian of the system. The (Cauchy) characteristic (ordinary) differential equations of the partial differential equations are the Hamilton equations. The "duality" between the partial and ordinary differential equations is the mathematical equivalent of the "wave-particle" duality of optics. It also played a key role (in the hands of de Broglie and Schrödinger) in the initial formulation of quantum mechanics.

Jacobi's theorem should also be mentioned here. If

$$z(x,\lambda)$$

is a complete family of solutions of the Hamilton equations, introduce

$$p_i = \frac{\partial z}{\partial \lambda^i} ,$$

so that

$$(p_i, \lambda^j)$$

form a new set of variables. In these variables, the Hamilton-Jacobi and Hamilton equations take their "canonical" form. (See DGCV, for example.) Finding such complete families is a very important and practical method for solving classical mechanics problems. (See Goldstein [1].) It is also a starting point for classical "perturbation" methods in celestial mechanics.

Returning to the case of a general first order partial differential equation (with one independent variable) we might remark that there is an extensive 19-th century theory (apparently going back at least to Lagrange), about the relation between the "complete solution", "general solution", "singular solution", etc. Very little of this has been passed down to modern mathematics. Treatises on partial differential equations, like Courant-Hilbert,

Vol. II, usually give some hurried description, but it is obviously only an anemic version of the rich 19-th century literature. The best classical treatises are Goursat and Forythes' books, although, of course, their bulk and archaic terminology is intimidating to the modern reader. E. Cartan wrote a long article "Sur certaines expressions differentielles et le problème de Pfaff", Part II, Vol. 1 of his Oeuvres, which applies his Pfaffian system formalism. It sums up much of the 19-th century literature, and is (for Cartan) relatively readable.

Cartan's main point is to show how much of the classical theory takes a very natural geometrical form in terms of Pfaffian systems. Following his point, we now reformulate the notion of "complete families of solutions" in terms of the theory of exterior differential systems.

3. COMPLETE FAMILIES OF SOLUTION MANIFOLDS OF EXTERIOR SYSTEMS

Keep the notation of Section 2. $M^1(X,Z)$ has the Lie coordinates
$$(x^i, z, z_j) \quad .$$

Set:
$$\theta = dz - z_i\, dx^i \tag{3.1}$$

θ is called the <u>contact form</u>. If $\underline{z} \in M(X,Z)$, then its prolongation $\partial^1\underline{z}$ is a solution submanifold of the Pfaffian system defined by θ, i.e.,

$$(\partial^1\underline{z})^*(\theta) = 0 \tag{3.2}$$

Conversely, an n-dimensional submanifold of

$$M^1(X,Z)$$

which is a solution of "$\theta = 0$", and <u>which is transversal to the variables</u> (x^i), is the prolongation of a $\underline{z} \in M(X,Z)$.

Suppose given a function

$$f: M^1(X,Z) \to R ,$$

(such that $df \neq 0$) and

$$DE \subset M^1(X,Z)$$

the submanifold on which f is zero.

Restrict θ and the functions (x^i) to DE. Let us give the submanifold DE the label

$$Y .$$

Let $P \subset F^1(Y)$ denote the $F(Y)$-submodule of one-forms on $F^1(Y)$ spanned by θ. P is a <u>Pfaffian system</u>.

COMPLETE FAMILIES

Definition. A <u>solution submanifold transversal to</u> (x^i) is an n-dimensional submanifold of Y such that:

θ restricted to the submanifold is zero
$dx^1 \wedge \cdots \wedge dx^n$ restricted to the submanifold is non-zero.

Retracing our steps, we see that giving such a submanifold is (at least locally) equivalent to giving a solution

$$x \to \underline{z}(x)$$

in the classical sense of the differential equation:

$$f(x, z(x), \partial z(x)) = 0 \ .$$

Having formulated in this Pfaffian system-geometric way what it means to solve the partial differential equation, we can now immediately formulate how a <u>complete family of solutions</u> would be defined.

Definition. Let Y be an m-dimensional manifold, with a Pfaffian system P, and with a set

$$(x^1, \ldots, x^n)$$

of functionally independent functions on Y. A <u>complete family of solutions</u> is defined by the following data:

a) A manifold Λ of dimension $(m-n)$.
b) A submersion map

$\pi: Y \to \Lambda$

such that:

c) For each $\lambda \in \Lambda$, the fiber $\pi^{-1}(\lambda)$ is an n-dimensional submanifold of Y which is a solution manifold of P, and which is transversal to $(x^1, \ldots, x^n)$.

Remark. In this formulation, there seems to be a close relation to Vessiot's approach to the theory of Pfaffian systems [2]. (See also the brief treatment in GPS.) He replaces Cartan's existence theorem with one which provides foliations, whose fibers are solution manifolds.

Chapter V

SOME GENERAL IDEAS OF GALOIS THEORY

1. INTRODUCTION

In pure mathematics, Galois theory is important mainly for historical reasons, since it is the source of much of modern algebra, e.g., the theory of groups, fields, etc. However, from the point of view of applications it still retains much more potential interest, because of its connection with solving certain types of problems by algorithms. In particular, Galois theory should be of interest in physics, engineering, systems theory, etc. because of its emphasis on the relation between the symmetries of a system and the possible algorithms for finding solutions of the system.

In the late 19-th century there was a lively interest--for example, by Picard, Vessiot, E. Cartan--in developing sufficiently general versions of the classical Galois theory to apply to broad classes of differential equations (and hence, also to physical systems). Much of this work is now lost to the scientific world, because of the way that mathematical notation and ways of thinking have changed. What survives is a small part of the whole, which deals with linear, ordinary differential equations, and which is now called "the" Picard-Vessiot theory. (See Kaplansky [1] and Kolchin [1].) It has survived because it is the only part that was stateable in terms of precise theorems, in the fashion of present-day mathematics. For me, the interest in Galois theory lies more in the vague, general ideas which motivated the 19-th century geometers, particularly the relation between Lie group theory and differential equations, and less the relatively uninteresting body of mathematics which is now given that name. In particular, one of my main goals is to understand the Galois approach to the notion of "symmetries" of differential or algebraic equations and the Lie approach. One is algebra, the other is geometry, but they are interrelated.

2. THE GENERAL IDEA OF GALOIS THEORY

Let A be a set, with some sort of mathematical structure S_A on it. This may be a structure of algebraic or geometric nature, perhaps an algebra, field, group or perhaps A is a manifold with a system of differential equations defined on it. Suppose also, given a group $G(A)$ of automorphisms of the structure S_A.

Suppose that a family $\{B\}$ of subsets of A is given such that the structure S restricts to define a

GALOIS THEORY

structure S_B on each subset B belonging to the family.

For each B of the family, let

$G(A,B)$ = subgroup of $g \in G(A)$ such that $g(b) = b$ for all $b \in B$.

Conversely, if H is a subgroup of $G(A)$, let:

$B_H = \{a \in A: h(a) = a \text{ for all } h \in H\}$.

Thus, we have mappings

$B \to G(A,B)$

$H \to B_H$

from sets to groups, and from groups to sets. <u>Galois theory</u> deals with this correspondence.

One could formalize this better in terms of categories and functions, for example, a category of sets and structures, and a category of groups, with the correspondence defining cofunctors between them. However, at the moment, I do not want to go into this.

Let us consider the simplest example, where A is a set with no particular structure, and where G is a given transformation group on A. For each subgroup

$H \subset G$,

set:

$$FS(H) = \{a \varepsilon A: ha = a \text{ for all } h \varepsilon H\}.$$

Consider the mapping

$$H \to FS(H)$$

from the set of subgroups of G to the set of subsets of A. Here are some properties:

If $H_1 \subset H_2$, then

$$FS(H_2) \subset FS(H_1) \tag{2.1}$$

If H is an invariant subgroup of G, then

$$GFS(H) \subset FS(H) \tag{2.2}$$

<u>Proof</u>. To say that H is an invariant subgroup of G is to say that

$$gHg^{-1} = H \tag{2.3}$$

for all $g \varepsilon G$.

For $a \varepsilon FS(H)$, $g \varepsilon G$, $h \varepsilon H$,

$$hga = g(g^{-1}hg)a$$

$$= ga, \text{ by (2.3)},$$

i.e., $ga \varepsilon FS(H)$.

This proves (2.2).

This result can be generalized.

Definition. If H is a subgroup of G, let

$$N(H,G) = \{g \varepsilon G: gHg^{-1} = H\} .$$

$N(H,G)$ is a subgroup of G, called the <u>normalizer of</u> H <u>in</u> G. It follows readily from its definition that:

H is an invariant subgroup of $N(H,G)$. (2.4)

(In fact, $N(H,G)$ is the <u>maximal</u> subgroup of G which contains H, in which H is an invariant subgroup.)

Now, we have:

$$N(H,G)(FS(H)) \subset FS(H) . \qquad (2.5)$$

Since H is an invariant subgroup of $N(H,G)$, the coset space

$$N(H,G)/H \qquad (2.6)$$

is a group called the <u>quotient group</u>. It acts as a transformation group on

$$FS(H) .$$

For, $h \varepsilon H$ <u>by definition</u> acts as the identity on $FS(H)$, hence, the action of $N(H,G)$ passes to the quotient to define an action of $N(H,G)/H$ on $FS(H)$.

<u>Exercise</u>. Here is the underlying fact. Suppose a group acts as a transformation group on a set B. Suppose

H is an invariant subgroup of L such that:

$$hb = b \text{ for all } h \in H .$$

Show that the quotient group L/H acts on B so as to lead to a commutative diagram:

$$\begin{array}{ccc} L \times B & \longrightarrow & B \\ \downarrow \uparrow & \downarrow \uparrow & \uparrow \downarrow \\ L/H \times B & \longrightarrow & B \end{array}$$

Now, we will describe an abstract "Galois theory".

Definition. A subset B of A is a <u>Galois subset of</u> A (relative to G) if there is a subgroup

$$H \subset G$$

such that:

$$B = FS(H) .$$

If B is a subset of A, set:

$$FG(B) = \{g \in G : gb = b \text{ for all } b \in B\} . \quad (2.7)$$

$FG(B)$ is called the <u>maximal subgroup of</u> G <u>which leaves</u> B <u>pointwise fixed</u>.

Also, set:

$$NG(B) = \{g \in G : gB \subset B\} \quad (2.8)$$

GALOIS THEORY

NG(B) is called the normalizing subgroup of B.

Exercise. Show that

FG(B) is an invariant subgroup of NG(B)   (2.9)

Definition. If B is a Galois subset of A, the quotient group

NG(B)/FG(B)

is called the Galois group of B, and denoted by

GG(B) .

By the exercise stated above, GG(B) acts as a transformation group on B.

Now, by an abstract Galois Theory we mean the process of:
   a) Picking out the family of Galois subsets B of A, defined by the transformation group action of G on A.
   b) Assigning the triple of groups

(FG(B), NG(B), GG(B) = NG(B)/FG(B))

to each Galois subset B of A.

Of course, further interesting study requires special assumptions about the "structure" assumed for A and G. The <u>classical Galois theory of algebraic equations</u> involves, in its modern formulation, the case where A is a field and G is a finite group of automorphisms of A. <u>The Galois theory of linear, ordinary differential equations</u> (usually called the <u>Picard-Vessiot Theory</u>) involves A as a real vector space, G as a Lie group which acts in an "algebraic" way on A.

Here is an even simpler example.

<u>Example</u>. <u>Linear subspaces of a finite dimensional vector space</u>

Let A be a finite dimensional vector space, with a field K as scalar field. Let G be the group of <u>all</u> invertible linear maps

$$g: V \to V \quad,$$

which is usually denoted by:

$$GL(V) \quad.$$

If H is a subgroup of GL(V), its fixed set is obviously a linear subspace of V. Conversely, each linear subspace W is the fixed set for <u>some</u> subgroup of GL(V). (To see this, choose any linear subspace W' so that $V = W \oplus W$. Then, GL(W') can be considered as a subgroup of GL(V), leaving fixed each element of W.

GALOIS THEORY

Clearly, W is the fixed set of this subgroup.) Hence:

The linear subspaces of V are the Galois subsets.

Given a linear subspace W, i.e., a Galois subset, note that:

$FG(W) = \{g \in GL(V): gw = w \text{ for all } w \in W\}$

$NG(W) = \{g \in GL(V): gW = W\}$.

Exercise. Show that $GG(W) = NG(W)/FG(W)$ is isomorphic to $GL(W)$.

In this Example, we see that the abstract Galois theory sketched above correctly predicts the main algebraic "structure" properties of finite dimensional vector spaces. Here is another feature.

Exercise. Show that the maps

$W \to FG(W)$

$W \to NG(W)$

are one-one maps from the set of linear subspaces of V to the set of subgroups of $GL(V)$. Further, the coset spaces

$GL(W)/NG(W)$

are important geometric objects--they are the <u>Grassman manifolds</u>.

In summary, the Galois ideas provide an abstract framework for investigating different geometric, algebraic and group-theoretic situations, and for developing <u>analogues</u> (even of an interdisciplinary type) between different mathematical situations.

<u>Remark</u>. One may regard the abstract Galois theory described above as a functor which maps a certain category of subsets of a set A into a category of exact sequences of groups. The functor maps the Galois set B into the exact sequence

$$(e) \to FG(B) \to NG(B) \to GG(B) \to (e)$$

Here is another general situation to apply Galois ideas. Let G be a group (e.g., a Lie group), S a subgroup (e.g., a closed Lie subgroup).

$$A = G/S$$

the coset space. One may now proceed to study the Galois situation.

It is not really necessary that G be a group to have all the abstract ideas <u>make sense</u>. For example,

GALOIS THEORY   101

G a <u>semigroup</u> (i.e., G have a multiplication satisfying the associative law) would be a typical generalization. Even non-associative situations, e.g., G = the Cayley algebra, might be an interesting generalization.

3. THE GALOIS THEORY OF FIELD EXTENSIONS AND POLYNOMIALS

Let us turn now to the situation which is usually called "Galois Theory" in the mathematical literature. (See Artin [1], Stewart [1].) In order to be able to prove the main facts, we shall have to derive some results from field theory.

<u>Theorem 3.1</u>. Suppose K and L are fields, and

$$\lambda_1,\ldots,\lambda_n: K \to L$$

are distinct, non-zero, one-one homomorphisms. Then, they are linearly independent over L.

<u>Proof</u>. Suppose that they are linearly dependent. This means that there exist elements $y_1,\ldots,y_n \in L$ (not all zero) such that:

$$y_1\lambda_1 + \cdots + y_n\lambda_n = 0 \tag{3.1}$$

In (3.1), some of the $(y_1,\ldots,y_n)$ may be zero. However, we may throw these away, to obtain a new expression of the same form, i.e., which satisfies (3.1), with all the $y_1,\ldots,y_n$ as non-zero elements of L.

We shall now take the relation (3.1), <u>with all the y's non-zero</u>, and convert it into a similar relation, but with a <u>smaller</u> n. Now, (3.1) means that

$$y_1\lambda_1(x) + \cdots + y_n\lambda_n(x) = 0 \tag{3.2}$$

for <u>all</u> $x \in K$.

Hence, for $x, x_1 \in K$,

$$y_1\lambda_1(xx_1) + \cdots + y_n\lambda_n(xx_1) = 0 \; .$$

But, $\lambda_1,\ldots,\lambda_n$ are <u>homomorphisms</u>. Hence,

$$y_1\lambda_1(x)\lambda_1(x_1) + \cdots + y_n\lambda_n(x)\lambda_n(x_1) = 0 \tag{3.3}$$

Multiply (3.2) by $\lambda_1(x_1)$, and subtract (3.3):

$$\begin{aligned}0 &= y_1\lambda_1(x)\lambda_1(x_1) + \cdots + y_n\lambda_n(x)\lambda_1(x_1) \\ & \quad - y_1\lambda_1(x)\lambda_1(x_1) - \cdots - y_n\lambda_n(x)\lambda_n(x_1) \\ &= (y_2\lambda_1(x_1) - y_2\lambda_2(x_1))\lambda_2(x) + \cdots \\ & \quad + (y_n\lambda_1(x_1) - y_n\lambda_n(x_1))\lambda_n(x)\end{aligned}$$

<u>Since the</u> $\lambda_1,\ldots,\lambda_n$ <u>are all distinct</u>, we can choose $x_1 \in K$ so that:

$$\lambda_1(x_1) \neq \lambda_2(x_1)$$

GALOIS THEORY

Thus, we have started with relation (3.2), with a given n, and derived a new relation of the same form, but with smaller n. Continuing, we arrive finally at a relation of form (3.2) with n = 1, which forces $\lambda_1 = 0$, which is the contradiction, and the proof of Theorem 3.1 is complete.

Theorem 3.2. Let G be a finite group of automorphisms of a field L. Let K be the fixed field of G, i.e.,

$$K = \{x \varepsilon L: g(x) = x \text{ for all } x \varepsilon L\} .$$

Then, the number of elements in G is equal to the dimension of L as a vector space over K.

Proof. Let $g_1, \ldots, g_n$ be the elements of G. Let $x_1, \ldots, x_m$ be elements of L which form a basis for L as a K-vector space. We may suppose that:

$$x_1 = 1 .$$

Then, each element $x \varepsilon L$ may be written in a unique way in the form:

$$x = k^1 x_1 + \cdots + k^m x_m ,$$
$$\text{with } k^1, \ldots, k^m \varepsilon K .$$

Let $y^1, \ldots, y^n \varepsilon L$. Set:

$$\lambda = y^1 g_1 + \cdots + y^n g_n .$$

Then, $\lambda = 0$ if and only if:

$$\lambda(x_a) = 0$$

for $1 \leq a \leq m$.

But,

$$\lambda(x_a) = y^1 g_1(x_a) + \cdots + y^n g_n(x_a) .$$

Hence, $\lambda(x_a) = 0$ if and only if the column vector

$$y = \begin{pmatrix} y^1 \\ \vdots \\ y^n \end{pmatrix}$$

satisfies:

$$Ay = 0 , \qquad (3.4)$$

where $A = (g_i(x_a))$ is an $m \times n$ matrix of elements of L.

Now, if

$$m < n ,$$

(3.4) always has a non-zero solution in y. (A set of $(n-1)$ homogeneous equations in n unknowns always has a non-zero solution in <u>a field</u>.) This would contradict Theorem 3.1, since the isomorphism $g_1, \ldots, g_n$ would be linearly dependent. Hence, we must have:

$$m \geq n ,$$

i.e.,

$$\text{order } (G) \leq \dim (L/K) . \qquad (3.5)$$

GALOIS THEORY   105

To finish the proof we must show that the equality sign holds in (3.5). Suppose otherwise, i.e.,

$$n < m.$$

Let

$$x_1, \ldots, x_{n+1}$$

be elements of L which are linearly independent over K. Using again the fact that n homogeneous equations in (n+1) unknowns have a non-trivial solution, we see that there are elements

$$z_1, \ldots, z_{n+1} \; \varepsilon \; L$$

such that:

$$\begin{aligned} g_1(x_1)z_1 + \cdots + g_1(x_{n+1})z_{n+1} &= 0 \\ &\vdots \\ g_n(x_1)z_1 + \cdots + g_n(x_{n+1})z_{n+1} &= 0 \end{aligned} \quad (3.6)$$

Let $g \; \varepsilon \; G$ act on both sides of (3.6). The result is:

$$\begin{aligned} (gg_1)(x_1)g(z_1) + \cdots + (gg_1)(x_{n+1})g(z_{n+1}) &= 0 \\ &\vdots \\ (gg_n)(x_1)g(z_1) + \cdots + (gg_n)(x_{n+1})g(z_{n+1}) &= 0 \end{aligned} \quad (3.7)$$

Now,

$$gg_1, \ldots, gg_n$$

is another listing of the elements of G. For example, suppose that:

$$gg_1 = g_2, \ldots, gg_n = g_1 .$$

Then, (3.7) reads as follows:

$$g_2(x_1)g(z_1) + \cdots + g_2(x_{n+1})g(z_{n+1}) = 0$$
$$\vdots \qquad\qquad\qquad\qquad (3.8)$$
$$g_1(x_1)g(z_1) + \cdots + g_1(x_{n+1})g(z_{n+1}) = 0$$

Multiply both sides of (3.6) by $g(z_1)$, and both sides of (3.8) by z_1. Subtract corresponding terms, leading to a relation of the form:

$$g_i(x_2)z_2' + \cdots + g_i(x_n)z_n' = 0$$

for $1 \le i, j \le n$.

Continuing in this way, we reduce the relation (3.6) down further and further, ultimately showing that it was impossible, i.e.,

$$m = n .$$

(See Stewart [1], p. 102, for the full details.)

<u>Theorem 3.3</u>. Let G be a finite group of automorphisms of a field L, with

$$K \subset L$$

the subfield of L which is the fixed set of G. For each subgroup $H \subset G$, set:

$$M_H = FS(H) .$$

GALOIS THEORY   107

Then, the assignment

$$H \to M_H$$

from (subgroups of G) → (subfields of L) is one-one.

<u>Proof</u>. Let H, H_1 be two subgroups which have the same fixed field $M \subset L$. By Theorem 3.2,

Set:
$$\text{order } H = \text{order } H_1 = L/M \ .$$

$$H_2 = \text{subgroup of } G \text{ generated by } H, H_1.$$

Note that:
$$M = \text{fixed set of } H_2.$$

Hence,
$$\text{order } H_2 = \text{order } L/M = \text{order } H = \text{order } H_1 \ .$$

Now, H is a subgroup of H_2. Since both are finite groups, they can have the same order only if:

$$H = H_2 \ .$$

Similarly,
$$H_1 = H_2 \ .$$

Hence, also
$$H = H_1 \ ,$$

which proves that the assignment

$$H \to \text{fixed field of } H$$

is indeed one-one, hence Theorem 3.3 is proved.

Now, we must show that each subfield is characterized by the subgroup which leaves it fixed. I will present the main idea involved in this, leaving details as an exercise, or to be consulted in Artin [1] or Stewart [1].

Let L be a field, K a subfield, with:

$$\dim (L/K) < \infty .$$

Let G = group of automorphisms which leave each element of K fixed. We want to find conditions which assure that conversely,

$$K = \text{fixed set for } G .$$

To do this, we must show that, for each

$$y \in L - K ,$$

there is a $g \in G$ such that

$$g(y) \neq y .$$

Now, our hypothesis is that L is finite dimensional as a K-vector space. Hence, the elements

$$1, y, y^2, \ldots, y^n$$

are linearly dependent (over K) for n sufficiently large. Introduce

$$K[x] ,$$

the polynomials with coefficients in K, in the "indeterminant" x. Hence, there is at least one $f \in K[x]$ such that:

GALOIS THEORY

$$f(y) = 0 . \qquad (3.9)$$

Choose f to be the polynomial of <u>minimal degree</u> satisfying (3.9). It is called the <u>minimal polynomial</u> of y.

Note that the minimal polynomial is <u>irreducible</u> as an element of K[x]. For, if $f_1 f_2 \in K[x]$ are lower degree polynomials, with

$$f = f_1 f_2 ,$$

then (3.9) implies that

$$f_1(y) = 0 \quad \text{or} \quad f_2(y) = 0 ,$$

which contradicts the minimal choice of f.

Let

K(y)

denote the smallest subfield of L containing y. It can be identified with the space of all values

$$\frac{p(y)}{q(y)} \qquad (3.10)$$

where $p, q \in K[x]$, $q(y) \neq 0$.

Let $z \in L$ be another element of L such that:

$$f(z) = 0 .$$

Then, since f is an irreducible element of K[x], it is <u>also</u> the minimal polynomial of z. Hence, we can define a map

$$K(y) \to K(z) \qquad (3.11)$$

by assigning to each element of $K(y)$ of form (3.10) the element

$$\frac{p(z)}{q(z)}$$

Exercise. Show that there is a well-defined map with these properties, and that it is a field isomorphism.

Now, both $K(y)$ and $K(z)$ are subfields of L. What are the conditions that the isomorphism (3.11) may be extended to an isomorphism:

$$L \to L \ ?$$

(If we succeed in finding such an isomorphism, it will have the property we need for our Galois theory, that it moves the given element $y \in L$.) To define this extension, proceed in a similar way. Let

$$y_1 \in L - K(y) \ .$$

Choose $f_1 \in K(y)[x]$ as the minimal polynomial of y_1 over $K(y)$. Let

$$f_2 \in K(z)[x]$$

be the polynomial which corresponds under the map (3.11). Let $z_1 \in L$ be a zero of f_2. One can then define an isomorphism

GALOIS THEORY   111

$$K(y)(y_1) \to K(z)(z_1) \quad,$$

extending the isomorphism (3.11).

    Suppose that all the elements $y, y_1, \ldots, z_1, z, \ldots$ needed for this construction lie in L. Since

$$\dim (L/K) < \infty \quad,$$

eventually our extension process must end with an isomorphism

$$L \to L$$

such that:

    Each $k \in K$ is fixed

    $y \to z$.

    When will this condition be satisfied? Clearly, if L is a "normal extension" of K, in the sense of the following:

Definition. Let L be a field, K a subfield. Then, L is a <u>normal extension</u> of K if the following condition is satisfied:

> If $f \in K[x]$ is a polynomial which has one root in L, then all of its roots lie in L, i.e., f factors into linear factors in L[x].   (3.12)

There are simpler ways of characterizing normality:

Exercise. Suppose that

$$\dim (L/K) < \infty .$$

Show that L is a normal extension of K if and only if there is a polynomial

$$f \in K[x]$$

of n-th degree, with roots

$$y_1,\ldots,y_n ,$$

such that:

$$L = K(y_1,\ldots,y_n) .$$

(In other words, L is obtained from K by adjoining **all** the roots of a polynomial.)

Here is the definitive result for Galois theory.

Theorem 3.4. Suppose that K is a field of characteristic zero, and that L is a field containing K, such that

$$\dim (L/K) < \infty .$$

Let G be the group of automorphisms of L which leave K fixed. Then, L is a normal extension of K if and only if K is the fixed field of G, i.e., for each $y \in L-K$, there is a $g \in G$ such that

$$g(y) \neq y .$$

GALOIS THEORY

Exercise. Prove Theorem 3.4. (See Artin [1], Stewart [1].) Note that the hypothesis of "characteristic zero" for K may be weakened to the condition that "L is a separable extension of K", where separability roughly means either that the characteristic of K is zero, or that L results from K by adjoining roots of polynomials whose derivative is not identically zero. "Separability" is a standard hypothesis in the algebraic geometry of fields of non-zero characteristic. With it, results proved for characteristic zero usually generalize.

We can now sum up the modern field-theoretic approach to the classical Galois Theory of algebraic equations. (We present the older ideas in the next section.)

Let K be a field, say, of characteristic zero. Let

$$f \in K[x]$$

be a polynomial with coefficients in K. Suppose that

$$y_1, \ldots, y_n$$

are the roots of f, and that

$$L = K(y_1, \ldots, y_n) \quad ,$$

i.e., L consists of the rational combinations of the y's, with coefficients chosen from K.

Let:

G = group of automorphisms of L which leave K fixed.

G is called the <u>Galois group of</u> f.

Now,
$$f(y_i) = 0$$
for $1 \leq i \leq n$

Hence, for $g \in G$,
$$f(g(y_i)) = 0 \;,$$
i.e., if $y_1,\ldots,y_n$ are the roots in a certain order, then
$$g(y_1),\ldots,g(y_n)$$
are these same roots <u>in a different order</u>. Thus,

<u>The Galois group is realized as a transformation group on the roots of</u> $f(x) = 0$.

However, an arbitrary permutation of the roots does not necessarily result from such an element of G. There are certain conditions which are imposed by the condition that g <u>must leave each element of</u> K <u>fixed</u>. We shall discuss this in the next section.

GALOIS THEORY

Having <u>defined</u> the Galois group of f, we can use the structure of its subgroups to study the structure of f. For each subgroup $H \subset G$, let

$$K_H = \text{fixed fields of } H.$$

Putting together all the results alluded to in this section, one can prove that the map

$$H \to K_H$$

<u>sets up a one-one correspondence between subgroups of</u> G <u>and subfields of</u> L.

Now, an "algorithm" for the "solution" of $f(x) = 0$ usually involves constructing an ascending tower of subfields:

$$K \subset K_1 \subset \cdots \subset K_m \subset L \qquad (3.13)$$

This corresponds to a descending tower of subgroups of G:

$$G \supset G_1 \supset \cdots \supset G_m \supset (e) \ . \qquad (3.14)$$

Conditions that the tower (3.13) of subfields satisfy usually translate into <u>group-theoretic properties of the tower</u> (3.14) <u>of subgroups</u>. For example, "solution by radicals" in the classical sense corresponds to the following conditions:

$$\begin{aligned} &G_1, \ldots, G_m \text{ are invariant subgroups of } G \\ &G_1/G_2, \; G_2/G_3, \ldots \text{ are abelian groups} \end{aligned} \qquad (3.15)$$

In turn, standard group theory implies that a tower of subgroups satisfying (3.14) exists if and only if G is a solvable group. Thus, we see the reason for the most famous conclusions of Galois theory:

A polynomial equation

$$f(x) = 0$$

is solvable by radicals if and only if its Galois group is "solvable" in the sense of group-theory. In particular, if f is of degree n, and if the permutation group on n letters is solvable, then every such polynomial is solvable by radicals. If the permutation group on n letters is not solvable, there exist polynomials of degree n which are not solvable by radicals. (In fact the "generic" polynomial of degree n will not be solvable by radicals.)

To complete the "solvability by radicals" story, note that in a purely group theoretic way (see Volume I) one proves that the permutation group in n letters is solvable only

for $n = 1, 2, 3, 4$.

GALOIS THEORY

Hence, one obtains a magnificent, almost Olympian, perspective on one fragment of the history of mathematics--explicit formulas were found for solvability by radicals of polynomials of degree 2, 3 and 4, but hundreds of years of effort failed to find a formula for fifth degree polynomials. Unfortunately, the Galois group is difficult to compute, so one cannot claim that this Galois Theory is very useful for the practical problem of finding the roots of algebraic equations. Still, I think of this theory as a model of "Applied Mathematics" of the highest sort.

It is interesting also to evaluate the relation of "modern mathematics" to the classical work. Its main contribution was making the theory understandable and "natural". By interpreting things in terms of standard algebraic concepts like fields and groups, the complicated, almost metaphysical, classical proofs became accessible, and, in fact, motivated much of the later development of algebraic geometry and algebraic number theory. Thus, the Galois ideas led to two "break throughs"--the first in Galois' original work, the second when mathematicians discovered the most natural algebraic setting for the ideas. There was then a further "resonance" of these break throughs in extension of these ideas to differential equations, first by Picard and Vessiot, in the difficult classical way, then in the modernization by Kolchin and Kaplansky, with the "natural" tie-in to the theory of algebraic groups.

4. THE CLASSICAL APPROACH TO GALOIS THEORY

As preparation for the extension to a Galois theory for differential equations, I will now briefly try to explain the classical approach, where the Galois group is defined as a certain subgroup of the permutation group.

Let

$$f \in C[x]$$

be a polynomial with complex coefficients. (Of course, in a "modern" approach, C could be replaced by an algebraically closed field whose characteristic (if non-zero) was such that the derivative of f was non-zero.) Suppose f is of the following form:

$$f(x) = x^n + a_{n-1}x^{n-1} + \cdots + a_0 \qquad (4.1)$$

Let K be the subfield of C containing the coefficients

$$a_0, \ldots, a_{n-1} .$$

Let $y_1, \ldots, y_n \in C$ be the roots of $f(x) = 0$. Then,

$$f(x) = (x-y_1) \cdots (x-y_n) \qquad (4.2)$$

GALOIS THEORY

Let $\Phi(f,K)$ be the set of polynomial maps

$$\phi: C^n \to C \qquad (4.3)$$

such that:

$$\phi(y_1,\ldots,y_n) \in K . \qquad (4.4)$$

There are elements ϕ in $\Phi(f,K)$.

Exercise. Show that each map (4.3) which is <u>symmetric</u> under permutation of its argument belongs to $\Phi(f,K)$. (This means that

$$\phi(z_{\pi(1)},\ldots,z_{\pi(n)}) = \phi(z_1,\ldots,z_n)$$

for each $z = (z_1,\ldots,z_n) \in C^n$,

each permutation $\pi: \{1,\ldots,n\} \to \{1,\ldots,n\}$.)

<u>Definition</u>. Let G be the group of one-one, onto maps

$$g: C^n \to C^n$$

such that:

$$\phi(g(z)) = \phi(z) \qquad (4.5)$$

for all $z \in C^n$, $\phi \in \Phi(f,K)$.

G is called the <u>Galois group of the polynomial equation</u>

$$f(x) = 0$$

Exercise. Let
$$L = K(z_1, \ldots, z_n)$$
Show that G may be identified with the group of automorphisms $: L \to L$ which leave each element of K fixed.

Here is an alternative way of defining G. Let
$$S(f) = \text{set of solutions of } f(x) = 0.$$
Define G as the set of invertible maps
$$\alpha: S(f) \to S(f)$$
such that:
$$\phi(\alpha(z_1), \ldots, \alpha(z_n)) = \phi(z_1, \ldots, z_n) \qquad (4.6)$$
for all $\phi \in \Phi(f, K)$.

This way of defining the Galois group is of more than historical interest. For, it should be clear (by my intuition, at least) that for a "generic" f, the *only* map in $\Phi(f, K)$ are those which are symmetric. This would imply that G consists of the *entire* permutation group on n letters, which is, as we have seen, tied to the possibility of "solvability by radicals". However, the real payoff for this approach comes in the understanding it gives us into a possible Galois Theory of Differential Equations.

GALOIS THEORY

5. THE GALOIS-PICARD-VESSIOT THEORY OF ORDINARY, LINEAR DIFFERENTIAL EQUATIONS

Consider a linear, homogeneous ordinary differential equation, say of second order, of the following form:

$$y'' + ay' + by = 0 \ . \qquad (5.1)$$

Here, $y(t)$ is an unknown C^∞, complex valued function of the real variable t, varying over some interval, say

$$a < t < b$$

a,b are C^∞ functions of t also on the same interval. $y'(t)$, $y''(t)$ denote the first and second derivatives. Let S be the space of all solutions of (5.1). The following facts are well-known:

S is a two-dimensional vector space over C, the complex numbers. Two functions $t \to y_1(t)$, $y_2(t)$ form a basis for S (as a C-vector space) if and only if their Wronskian

$$\begin{vmatrix} y_1 & y_2 \\ y_1' & y_2' \end{vmatrix}$$

is non-zero.

Suppose that y_1, y_2 are two functions which form a basis for S. We can solve for a,b in terms of y_1, y_2:

$$ay_1' + by_1 = -y_1''$$
$$ay_2' + by_2 = -y_2'' ,$$

or

$$a = \frac{\begin{vmatrix} -y_1'' & y_1 \\ -y_2'' & y_2 \end{vmatrix}}{\begin{vmatrix} y_1' & y_1 \\ y_2' & y_2 \end{vmatrix}} \tag{5.2}$$

$$b = \frac{\begin{vmatrix} y_1' & -y_1'' \\ y_2' & -y_2'' \end{vmatrix}}{\begin{vmatrix} y_1' & y_1 \\ y_2' & y_2 \end{vmatrix}} \tag{5.3}$$

We can rewrite (5.2)-(5.3) in the following form:

$$a = f_1(y_1, y_2, y_1', y_2', y_1'', y_2'')$$
$$b = f_2(y_1, y_2, y_1', y_2', y_1'', y_2'') \tag{5.4}$$

where f_1, f_2 are rational functions (in the sense of algebraic geometry) of the variables indicated by the formula. <u>Equations</u> (5.4) <u>constitute a system of differential equations for</u> y_1, y_2 <u>which are equivalent to finding the general solution of</u> (5.1).

GALOIS THEORY

Consider the group $GL(2,C)$ of 2×2 matrices, with constant coefficients. Denote a $g \in GL(2,C)$ by:

$$g = \begin{pmatrix} a_{11} & a_{12} \\ a_{21} & a_{22} \end{pmatrix}$$

Let g act on the variables

$$(y_1, y_2, y_1', y_2', y_1'', y_2'')$$

as follows:

$$g(y_1, y_2, y_1', y_2', y_1'', y_2'')$$
$$= (a_{11}y_1 + a_{12}y_2,\ a_{21}y_1 + a_{22}y_2,\ a_{11}y_1' + a_{12}y_2',\ a_{21}y_1' + a_{22}y_2', \ldots)$$

In other words, we identify the point

$$(y_1, y_2, y_1', y_2', y_1'', y_2'')$$

with

$$C^2 \oplus C^2 \oplus C^2,$$

and let $GL(2,C)$ act in the natural way $GL(2,C)$ acts on two-vectors.

Theorem 5.1. $GL(2,C)$ leaves the differential equations (5.4) invariant, in the sense that if (for given a,b), $y_1(t)$, $y_2(t)$ satisfies (5.4), then so does

$$g \begin{pmatrix} y_1 \\ y_2 \end{pmatrix},$$

where $g \in GL(2,C)$.

The proof should be obvious. Notice that here it is a question of <u>invariance of a system of differential equations in the sense meant by Sophus Lie</u>, i.e., a group (in this case GL(2,C)) acts on the manifold on which the differential equations are defined, and preserves the set of solutions.

Now, choose a fixed solution (y_1, y_2) of (5.2)-(5.3), i.e., a fixed basis of S. Let Ω be a fixed set of C^∞ functions of t, $0 < t < b$. In the version of Picard-Vessiot theory which is now standard, e.g., Kolchin [1], Kaplansky [1], Ω is a set of functions which form a <u>differential field</u>, i.e., is closed under addition, multiplication, inverses and derivatives. However, there is no particular <u>geometric</u> reason for making such an assumption, or for choosing the framework of "differential algebra" for that matter. (Since "differentiation" depends on the coordinates, the notion of "differential field automorphism" is not natural from the geometric point of view.) To have a <u>geometrically</u> satisfactory purely algebraic theory one should redo things on the model of the theory of <u>mapping element spaces</u>. In fact, the original version by Picard [1, Volume III] and Vessiot [1], which I am essentially following, is much better suited to the modern differential-geometric viewpoint, although it is more difficult to work with

GALOIS THEORY

technically than the Kolchin framework.

Having fixed Ω, let us say that a rational mapping (in the sense of algebraic geometry)

$$\phi: C^{2m} \to C$$

is Ω-<u>admissable</u> if:

$$(y_1, y_2, y_1', y_2', \ldots, y_1^{(m)}, y_2^{(m)}) \in \Omega \tag{5.5}$$

<u>Definition</u>. The <u>Galois-Picard-Vessiot group</u> of the differential equation (5.1), with respect to the set Ω, consists of the $g \in GL(2,C)$ which leave invariant <u>all</u> the Ω-admissible maps ϕ, for <u>all</u> integers m. Then we have:

$$\phi(g(z)) = \phi(z), \tag{5.6}$$

for all $z = (z_1, z_2, \ldots, z_{2m}) \in C^{2m}$,
all ϕ satisfying (5.5).

<u>Exercise</u>. Show that this condition is independent of the basis chosen for S, so that the Galois-Picard-Vessiot group can be defined directly as a group of linear maps : S $\to$ S.

<u>Remark</u>. Note further that (5.6) is an <u>algebraic</u> condition on $g \in GL(2,C)$, so that the Galois-Picard-Vessiot group is an <u>algebraic matrix group</u>. In turn, this fact has

played a very important role in the modern theory, since the theory of algebraic matrix groups is very well-developed. Of course, as a <u>closed</u> subgroup of $GL(2,C)$, it is also a <u>Lie group</u>.

To recapture the differential algebra viewpoint, let Ω be a differential field (containing the coefficients a,b of the original differential equation (5.1)), and let

$$\Omega\{y_1, y_2\}$$

be the smallest differential field containing Ω, y_1 and y_2. One identifies g in the Galois-Picard-Vessiot group with an automorphism (in the <u>differential algebra</u> sense) of $\Omega\{y_1, y_2\}$ which leaves Ω pointwise fixed. Conditions (5.5) and (5.6) guarantee that this can be done <u>consistently</u>. However, this framework is not as broad as the original one suggested by Picard and Vessiot, since it does not generalize easily to non-linear differential equations.

Now, let us briefly investigate the group-theoretic meaning of "solvability by quadratures".

GALOIS THEORY 127

6. SOLVABILITY BY QUADRATURES AND SOLVABLE LIE GROUPS

Continue to work with the second order, linear differential equation (5.1). (The methods generalize.) Let Ω be a set of functions of t, which include the coefficients

a,b in (5.1) .

<u>Definition</u>. The differential equation (5.1) is said to be <u>solvable by quadratures</u> (with respect to the function Ω) if there is a solution y_1 of (5.1) such that:

$$y_1' \in \Omega \ . \tag{6.1}$$

<u>Exercise</u>. If (5.2) is satisfied, show that there is a linearly independent pair (y_1, y_2) of solutions of (5.1) such that:

$$y_1' \in \Omega \tag{6.2}$$

y_2' can be expressed as rational
functions of elements in Ω, y_1 (6.3)
and y_1'.

<u>Remark</u>. If Ω is a differential field, and if $\Omega\{y_1, y_2\}$ is the differential field generated by y_1, y_2, conditions (6.2)-(6.3) mean that, in the terminology of Kaplansky [1],

$\Omega\{y_1,y_2\}$ results from Ω by successive <u>Liouville extensions</u>.

Now, suppose that the equation (5.1) is solvable by quadratures with respect to Ω, and that y_1, y_2 are a basis for the solution space S which satisfies (6.2).

Let ϕ be the linear map

$$C^4 \to C$$

defined as follows

$$\phi(z_1, z_2, z_3, z_4) = z_3 \ . \tag{6.4}$$
$$\text{for } z = (z_1, z_2, z_3, z_4) \in C^4$$

Note that condition (6.2) means that:

$$\phi(y_1, y_2, y_1', y_2') \in \Omega \ ,$$

i.e., ϕ satisfies (5.5), hence is a Ω-<u>admissible rational map</u>. (In fact, it is linear, of course.)

Let G be the Galois-Picard-Vessiot group, considered as a subgroup of $GL(2,C)$. If $g \in G$, then ϕ must preserve the map ϕ. This means that:

$$a_{21} = 0 \ ,$$

where

$$g = \begin{pmatrix} a_{11} & a_{12} \\ a_{21} & a_{22} \end{pmatrix}$$

GALOIS THEORY   129

This means that G is a solvable group, in the usual
group-theoretic sense. Conversely (at least in the
"differential algebra" context) solvability of G implies
solvability by quadratures, but this is obviously considerably more difficult to prove, since it requires sophisticated algebra-geometric tools.

Exercise. Take for Ω the field of rational functions
of t, and for the differential equation (5.1) one
which is integrable by quadratures with coefficients
$a,b \in \Omega$. (See any elementary differential equation text
for such examples.) In such an example, calculate a
basis for the admissible Ω-maps

$$\phi: C^m \to C ,$$

and calculate the Galois group.

7. REDUCIBILITY OF THE GALOIS-PICARD-VESSIOT GROUP AND REDUCIBILITY OF DIFFERENTIAL OPERATORS

In the last section we have seen that an algebraic
property of the Galois-Picard-Vessiot group of the differential equation (5.1)--solvability--is related to an
analytic property of the differential equation--solvability
by quadratures. Now, we provide another example--the
relation between reducibility of the group and reducibility
of the differential operator. (In fact, in this particular

case, reducibility is <u>equivalent</u> to integrability by quadratures, but we shall do things in such a way as to make possible the generalization to higher order equations where they are not equivalent.

Associate with equation (5.1) the differential operator:

$$D(y) = y'' + ay' + by .$$

If coefficients a, b lie in a set Ω of functions, we say that D <u>is defined over</u> Ω. (Of course, its solutions do not have to be in Ω.)

<u>Definition</u>. D <u>is reducible over</u> Ω if there are differential operators D_1, D_2 (of lower order) which are also defined over Ω such that:

$$D = D_1 D_2 \tag{7.1}$$

Suppose that D is reducible in this way: We shall show that this condition implies <u>reducibility</u> of the Galois-Picard-Vessiot group, as a subgroup of $GL(2,C)$.

To see this, note that we can find a basis (y_1, y_2) for the solutions of (5.1) such that:

$$D_2(y_1) = 0 . \tag{7.2}$$

Suppose that:

$$D_2(y) = a_2 y' + b_2 y ,$$

GALOIS THEORY 131

with $a_2, b_2 \in \Omega$. Then,

$$a_2 y_1' + b_2 y_1 = 0 ,$$

or

$$\frac{y_1'}{y} = -\frac{b_2}{a_2} \tag{7.3}$$

Let

$$\phi: C^4 \to C$$

be the rational map defined as follows:

$$(z_1, z_2, z_3, z_4) = \frac{z_3}{z_1} \tag{7.4}$$

Since $a_2, b_2 \in \Omega$, note that (supposing that elements of Ω have inverses which again lie in Ω) condition (7.3) means that:

$$\phi(y_1, y_2, y_1', y_2') \in \Omega \tag{7.5}$$

Thus, ϕ is an Ω-admissible rational map. The Galois group G must then preserve it. Then, if

$$g = \begin{pmatrix} a_{11} & a_{12} \\ a_{21} & a_{22} \end{pmatrix}$$

is an element of $GL(2,C)$, we must have:

$$\frac{a_{11} z_3 + a_{12} z_4}{a_{11} z_1 + a_{12} z_2} = \frac{z_3}{z_1} ,$$

or

$$a_{11}z_3z_1 + a_{12}z_4z_1 = a_{11}z_1z_3 + a_{12}z_3z_4 ,$$

or

$$a_{12} = 0 \qquad (7.6)$$

Again, we see that condition (7.2) forces G to act reducibly on C^2. (Of course, in this case, this is equivalent to solvability of G, but this will not be true for higher order equations.)

Exercise. Investigate directly (i.e., without directly using the Galois group) the relation between reducibility of the second order differential operator D and solvability-by-quadratures of the differential equation $Dy = 0$.

8. PRINCIPAL LIE SYSTEMS

In Volumes III and IX I have emphasized the role-- and the importance for practical applications--of the special sort of differential equations called Lie systems. Of course, this emphasis seems to be nothing new--but to have been very well known around the turn of the century-- see Vessiot's review [1]--but the ideas certainly seem to have disappeared from the current mathematical literature!

Now, Lie systems are differential equations associated with transformation group actions of Lie groups on

GALOIS THEORY

manifolds. An important special case is where the manifold is the Lie group itself, and the differential equations are invariant under left translation. These are called <u>principal Lie systems</u>. We shall now study such systems from the Galois-Picard-Vessiot viewpoint.

Let G be a Lie subgroup of GL(n,C). (Using differential geometry, the theory could readily be extended to abstractly defined Lie groups, but it is useful to keep as close as possible to the classical situation.) The <u>Lie algebra</u> of G, denoted by $\underset{\sim}{G}$, is also a set of n×n complex matrices. To obtain it, consider the C^∞ curves

$$t \to g(t)$$

in G. $\underset{\sim}{G}$ consists of the matrices a which can be written in the form:

$$\frac{dg(t)}{dt} g(t)^{-1} = a, \qquad (8.1)$$

for some t, some curve in G.

Conversely, one can prove (and it is a moderately deep result!) that if

$$t \to a(t)$$

is any curve (say, continuous), and if $t \to g(t)$ is a solution of (8.1) such that:

$$g(t_0) \in G$$

for one value of t, then:

$$g(t) \in G \quad \text{for } \underline{\text{all}} \; t \; .$$

Differential equation (8.1), usually written in the following form:

$$\frac{dg}{dt} = a(t)g(t) \; , \qquad (8.2)$$

is called a __principal Lie system__. Its solutions are invariant under right translation, in the following sense:

If $g_0 \in G$, and $t \to g(t)$ is a solution of (8.1), then

$$t \to g(t)g_0$$

is also a solution of (8.1).

__Example__. Consider the differential equation (5.1)

$$y'' + \alpha(t)y' + \beta(t)y = 0 \; . \qquad (8.4)$$

Let y_1, y_2 be two linearly independent solutions. Set:

$$g(t) = \begin{pmatrix} y_1(t) & y_2(t) \\ y_1'(t) & y_2'(t) \end{pmatrix}$$

It is a curve in $GL(2,C)$ which is a solution of a system of form (8.2), with a 2×2 matrix a constructed from $\alpha_1 \beta$.

Exercise. Find a in terms of $\alpha_1 \beta$.

The action (8.3) of $G = GL(2,C)$ on solutions then simply corresponds to the transformation

$$y_1 \to a_{11} y_1 + a_{12} y_2$$
$$y_2 \to a_{21} y_1 + a_{22} y_2 \qquad (8.5)$$

that we have already exploited to define the Galois-Picard-Vessiot group.

We can now generalize the way we have earlier defined the Galois group of the equation (5.1) to define it for Lie systems of this type. Again, it will be defined as the subgroup of G which preserves certain "rationality" relations. To define this in a simple way, it will be most convenient to define the prolonged groups of G to various orders.

9. THE PROLONGED GROUP OF A LIE GROUP

We continue to suppose that the Lie group G is a subgroup of $GL(n,C)$. Standard results of differential geometry would enable one to work "intrinsically", without this assumption, but the notation would be somewhat more complicated.

Let

$$\Gamma(R,G)$$

be the space of C^∞ maps

$$\gamma: R \to G \quad.$$

Let

$$A(R,G) \subset G \times \Gamma(R,G)$$

be the space of all pairs

$$(g,\gamma)$$

such that:

$$\gamma(0) = g \qquad (9.1)$$

Make $A(R,G)$ into a group by defining the multiplication in $A(R,G)$ via the following formula:

$$(g,\gamma)(g_1,\gamma_1) = (gg_1,\gamma\gamma_1) \qquad (9.2)$$

For each integer r, let:

$$A^r(R,G)$$

be the subset of

$$(g,\gamma) \in A(R,G)$$

such that:

$$g = 1, \quad \frac{d\gamma}{dt}(0) = 0 = \cdots = \frac{d^r\gamma}{dt}(0) \qquad (9.3)$$

GALOIS THEORY   137

Theorem 9.1. $A^r(R,G)$ is an invariant subgroup of $A(R,G)$.

Proof. First, we prove that $A^r(R,G)$ is a <u>subgroup</u>. Suppose

$$(g,\gamma), \; (g_1,\gamma_2) \; \varepsilon \; A^r(R,G) \; .$$

By (9.2), the product in $A(R,G)$ is given as follows:

$$(g,\gamma)(g_1,\gamma_1) = (gg_1,\gamma\gamma_1) \; .$$

Now, if γ and γ_1 both satisfy (9.3), so does their product. Hence, $A^r(R,G)$ is a subgroup.

Let us now show that is is an <u>invariant</u> subgroup. Let

$$(g,\gamma) \; \varepsilon \; A(R,G), \quad (g_1,\gamma_1) \; \varepsilon \; A^r(R,G) \; .$$

Then,

$$(g,\gamma)(g_1,\gamma_1)(g,\gamma)^{-1} = (g,\gamma)(g_1,\gamma_1)(g^{-1},\gamma^{-1})$$

$$= (gg_1 g^{-1}, \gamma\gamma_1\gamma^{-1})$$

$$= (1, \gamma\gamma_1\gamma^{-1}) \; ,$$

since $g_1 = 1$. (1 is the identity element of G.)

Now,

$$\frac{d}{dt}(\gamma\gamma_1\gamma^{-1}) = \left(\frac{d\gamma}{dt}\right)(\gamma_1)(\gamma^{-1}) + \frac{d\gamma_1}{dt}\gamma^{-1} - \gamma\gamma_1\gamma^{-1}\frac{d\gamma}{dt}\gamma^{-1}$$

Hence,

$$\frac{d}{dt}(\gamma\gamma_1\gamma^{-1})(0) = \frac{d\gamma}{dt}(0)\gamma(0)^{-1} + 0 - \gamma(0)1\gamma(0)^{-1}\frac{d\gamma}{dt}(0)\gamma(0)^{-}$$

$$= 0 \ .$$

Similarly, one proves that the first r derivatives of the curve

$$t \to \gamma\gamma_1\gamma^{-1}$$

vanish at $t = 0$, i.e.,

$$(g,\gamma)(g_1,\gamma_1)(g,\gamma)^{-1} \in A^r(R,G) \ ,$$

which finishes the proof of Theorem 9.1.

<u>Definition</u>. The r-th prolonged group of G, denoted by G^r, is defined as the following quotient group:

$$G^r = A(R,G)/A^r(R,G) \ . \tag{9.4}$$

<u>Theorem 9.2</u>. G^r is a Lie group. G^1 is diffeomorphic to $T(G)$, the tangent bundle to G.

<u>Proof</u>. Assign to $(g,\gamma) \in A(R,G)$ the following element:

$$\left(g, \frac{d}{dt}(\gamma)\gamma^{-1}(0), \left(\frac{d}{dt}\frac{d\gamma}{dt}\gamma^{-1}\right)(0), \ldots, \frac{d^{r-1}}{dt^{r-1}}\left(\frac{d\gamma}{dt}\gamma^{-1}\right)(0)\right)$$

(9.5)

of $G \times \underset{\sim}{G} \times \cdots \times \underset{\sim}{G}$.

This defines a map

$$\phi: A(R,G) \to G \times \underbrace{\underset{\sim}{G} \times \cdots \times \underset{\sim}{G}}_{r \text{ times}}$$

(9.6)

We shall show that ϕ passes to the quotient to define a map, also denoted by ϕ,

of $G^r \to G \times \underset{\sim}{G} \times \cdots \times \underset{\sim}{G}$.

To see this, suppose $(g_1, \gamma_1) \in A^r(R,G)$, $(g,\gamma) \in A(R,G)$. Then,

$$(g,\gamma)(g_1,\gamma_1) = (gg_1, \gamma\gamma_1)$$

Then,

$$\frac{d}{dt}(\gamma\gamma_1)(0) = \frac{d\gamma}{dt}\gamma_1(0) + \gamma(0)\frac{d\gamma_1}{dt}(0)$$

$$= \frac{d\gamma}{dt}(0),$$

since (g_1, γ_1) satisfies relations (9.3).

For $r = 1$, this shows that:

$$\phi((g,\gamma)A^r(R,G)) = \phi(g,\gamma)$$

(9.7)

The proof obviously generalizes, and (9.8) holds for
<u>all</u> r. This shows that ϕ passes to the quotient to
define a map

$$\phi: G^r \to G \times \underset{\sim}{G} \times \cdots \times \underset{\sim}{G} \qquad (9.8)$$

<u>Exercise</u>. Prove that the map (9.9) is one-one, and
onto. Prove that, with respect to the manifold structure
on G^r provided by ϕ, G^r is a <u>Lie group</u>, thus finishing the proof of Theorem 9.2.

We can easily calculate the group law on G^r in
terms of its parameterization, via ϕ, by $G \times \underset{\sim}{G} \times \cdots \times \underset{\sim}{G}$.
For example, let us carry out this calculation in case:

$$r = 1 \ .$$

Suppose that

$$(g,\gamma), \ (g_1,\gamma_1) \quad \varepsilon \quad A(R,G).$$

Then,

$$\frac{d}{dt}(\gamma\gamma_1)(\gamma\gamma_1)^{-1}(0) = \left(\frac{d\gamma}{dt}(0)\gamma_1(0)+\gamma(0)\frac{d\gamma_1}{dt}(0)\right)(\gamma_1(0)^{-1}\gamma(0)^{-1})$$

$$= \frac{d\gamma}{dt}(0)\gamma(0)^{-1}+\gamma(0)\frac{d\gamma_1}{dt}(0)\gamma_1(0)^{-1}\gamma(0)^{-1}$$

$$= \frac{d}{dt}(0)\gamma(0)^{-1}+g\left(\frac{d\gamma_1}{dt}(0)\gamma_1(0)^{-1}\right)g^{-1}$$

$$(9.9)$$

GALOIS THEORY

We can interpret this theorem in the following way:

Theorem 9.3. Regard G^1 as diffeomorphic to $G \times \underset{\sim}{G}$, as explained above. Then, the Lie group law inherited by $G \times \underset{\sim}{G}$ via this diffeomorphism is given by the following formula:

$$(g,X)(g_1,X_1) = (gg_1, X + \text{Ad } g(X_1)) \qquad (9.10)$$

$$\text{for } g, g_1 \in G; \quad X, X_1 \in \underset{\sim}{G} \, .$$

In turn, (9.10) may be interpreted by saying that G^1 is a semidirect product of G and an abelian group isomorphic to the abelian vector group of $\underset{\sim}{G}$, with the representation of G on the vector space $\underset{\sim}{G}$ defining the semidirect product just the <u>adjoint representation</u> of Lie theory.

<u>Exercise</u>. Generalize formula (9.10) to the case of general r.

<u>Remark</u>. Notice that the formula for the group law in G^1 <u>does not depend on the realization of</u> G <u>as a group of matrices</u>, but depends only on its Lie group structure. This is a confirmation of my claim that the concepts can be defined in an intrinsic way, but that it is convenient to use a matrix representation!

We have now done enough with these prolonged groups to formulate the Galois theory.

10. THE GALOIS-PICARD-VESSIOT GROUPS OF PRINCIPAL LIE SYSTEMS

Let us recapitulate notation. G is a Lie subgroup of $GL(n,C)$. $\underset{\sim}{G}$ denotes its Lie algebra.

$$t \to a(t)$$

is a curve in $\underset{\sim}{G}$. It defines the following Lie system:

$$\frac{dg}{dt} = ag \qquad (10.1)$$

Suppose that one solution of (10.1) is fixed. Let Ω denote a given set of functions of t, defined over the same interval as the solution of the differential equation (10.1) is defined. Let $G^{(r)}$, for $r = 1, 2, \ldots$, be the <u>prolonged group</u> of G, as defined in Section 9. Then, the curve

$$\gamma: t \to g(t)$$

in G defines a curve

$$t \to \partial^r g$$

in $\underset{\sim}{G}^r$.

In fact, if $t \in R$,

$$\partial^r g(t)$$

is the quotient in $G^r = A(R,G)/A^r(R,G)$ of the element

$$(g(t), \gamma)$$

GALOIS THEORY

in $A(R,G)$. For example, with the identification of
G^1 with $G \times \underset{\sim}{G}$
explained in Section 9,

$$\partial^1 g(t) = \left(g(t), \frac{dg}{dt} g(t)^{-1}\right) \tag{10.2}$$

Exercise. Prove (10.2). Find the generalization of this formula for $r = 2,3,\ldots$

Definition. A map

$$\alpha: G^r \to C$$

is Ω-<u>admissible</u> with respect to the given curve $\gamma: t \to g(t)$ if the following condition is satisfied:

The function

$$t \to \alpha(\partial^r g(t)) \tag{10.3}$$

lies in Ω

Right translation

$$g \to gg_0^{-1}$$

defines an action of G on G. This action "prolongs" to define a transformation group action

$$G \times G^r \to G^r \quad .$$

Definition. The Galois-Picard-Vessiot group of the principal Lie system defined by the curve $t \to a(t)$ in $\underset{\sim}{G}$ is the subgroup of G consisting of the elements $g_0 \in G$ which, acting on G^r as explained above, leave invariant the Ω-admissible maps α.

Remark. If $t \to g_1(t)$ is another solution of (10.1), then there is a $g_0 \in G$ such that

$$g(t) = g_1(t)g_0 ,$$

i.e., the two solutions of (10.1) differ by right translation under g_0. Hence, the set of Ω-admissible maps differ by right translation. This shows that the Galois-Picard-Vessiot subgroups of G associated with the two solutions of (10.1), (and the choice of Ω) are conjugate under an inner automorphism of G.

The admissible maps α are objects that Sophus Lie called differential invariants. Thus, the study of the Galois groups of differential equations is really (as Vessiot himself remarks [1]) a special situation in the theory of differential invariants.

If G is an algebraic group, so is G^r. Hence, one can restrict attention to the Ω admissible maps $\alpha: G^r \to C$ which are rational maps. In this way, it is possible to give the Galois group GG the structure of an algebraic group.

GALOIS THEORY

The next step is to develop the relation between the subgroups of the Galois group and the special techniques of solvability of the equation (10.1). For example, if the differential equations (10.1) are "solvable by quadratures", one sees readily--just as in the classical Picard-Vessiot situation--that the Galois group is _solvable_ in the Lie group sense. Presumably, the converse also holds, although there does not seem to be a proof, at least in the modern literature.

Chapter VI

LIE AND BÄCKLUND SYMMETRIES OF DIFFERENTIAL EQUATION SYSTEMS

1. INTRODUCTION

Lie was probably the first one in the 19-th century to systematically use differential forms and vector fields in the study of differential equations. His ideas along these lines were brought to at least a partial completion in a magnificent series of papers by Elie Cartan, ranging from 1899 to 1915, printed as papers 14 to 45 in Part II of his Collected Works. There has been very little progress since then.

Of course, there has, in recent years, been extensive and impressive work on systems of differential equations, but it has been more along the lines of pure algebra and analysis, rather than the "geometric" approach of Lie and Cartan, which I prefer and believe still have vast unexplored potentialities. The most notable and sustained work of this sort has been by D.C. Spencer and his collaborators. See Kumpera-Spencer [1].

My main task in this chapter is to set up some of the ideas in a more coordinate-free way, using the algebraic notion of "linear differential operator", developed purely algebraically in GPS. I will also deal with "generalized solutions" (which I call Lie solutions) and "contact transformations" (which I call Lie symmetries) of systems of differential equations.

For introductionary treatments of this material, see Cartan's book "Les systemes differentielles et leurs applications geometrique", my paper "E. Cartan's geometric theory of differential equations", and GPS.

2. MAPPING ELEMENT SPACES DEFINED IN TERMS OF LINEAR DIFFERENTIAL OPERATORS

In this chapter we work with the following assumptions:

X is a manifold of dimension n. A typical coordinate system for X is denoted as

$$(x^i), \quad 1 \leq i,j \leq n \ .$$

Z is a finite dimensional real vector space of dimension m. A typical coordinate system of Z is denoted by

$$(z^a), \quad 1 \leq a,b \leq m \ .$$

This coordinate system is not necessarily linear, although that is, of course, a reasonable choice to make.

SYMMETRIES   149

Let:

$F(X)$ = ring of C^∞, real-valued functions on X

$\Gamma(X,Z)$ = space of C^∞ maps $\gamma: X \to Z$.

Since Z is a vector space, elements of $\Gamma(X,Z)$ can be added, and multiplied by elements of $F(X)$, i.e.,

$\Gamma(X,Z)$ is an $F(X)$-module .

Thus, we can apply the purely algebraic definition of "linear differential operator", developed, in a module context, in GPS, Chapter I.

In particular, set:

$D(X,Z)$ = space of linear differential operators $\Gamma(X,Z) \to F(X)$.

Remark. In terms of the notation of GPS,

$$D(X,Z) \equiv D(\Gamma(X,Z), F(X)) \qquad (2.1)$$

Let:

Δ be a subset of $D(X,Z)$.

Introduce an equivalence relation into

$X \times \Gamma(X,Z)$

as follows:

(x,γ) is equivalent to (x',γ') if the following conditions are satisfied:

$$x = x'$$
$$\gamma(x) = \gamma(x') \qquad (2.2)$$
$$\delta(\gamma)(x) = \delta(\gamma')(x)$$

for all $\delta \in \Delta$

Let

$$M(\Delta,X,Z)$$

be the quotient of $X \times \Gamma(X,Z)$ under this equivalence relation. It is called the <u>mapping element space defined by</u> Δ.

<u>Exercise</u>. If Δ = set of <u>all</u> r-th order linear differential operators : $\Gamma(X,Z) \to F(X)$, show that

$$M(\Delta,X,Z) = M^r(X,Z) \qquad (2.3)$$

<u>Remark</u>. Thus, using the coordinate-free definition of "differential operator" we have generalized the classical notion of "mapping element space" (or "jet-space, in the sense of Ehresmann". I do not know if this generalization is significant, but it is useful to have a coordinate-free formalism. It would also be possible to replace the condition that Z be a vector space by the condition that it be a manifold, at the cost of a slightly more cumbersome definition.

SYMMETRIES

All the ideas developed earlier associated with the usual "mapping element spaces" generalize. Here are some examples:

If $\gamma \in \Gamma(X,Z)$, $x \in X$, let $\partial \gamma(x)$ denote the equivalence class to which $(x,\gamma) \in X \times \Gamma(X,Z)$ belongs. As x varies, we obtain a map

$$\partial \gamma: X \to M(\Delta, X, Z) \quad , \tag{2.4}$$

called the <u>partial derivative</u> of γ. A differential subset

$$S \subset M(\Delta, X, Z)$$

defines a <u>differential system</u>. A $\gamma \in \Gamma(X,Z)$ is a <u>solution</u> of S if:

$$\partial \gamma(X) \subset S \quad . \tag{2.5}$$

Each $\delta \in \Delta$ now defines a function:

$$h_\delta: M(\Delta, X, Z) \to R$$

as follows:

$$h_\delta(\partial \gamma(x)) = \delta(\gamma)(x) \quad . \tag{2.6}$$

<u>Remark</u>. Here is the meaning of this formula. Define a function

$$X \times \Gamma(X,Z) \to R$$

by the formula

$$(x,\gamma) \to \delta(\gamma)(x) \quad .$$

(Recall that δ is a linear differential operator $\Gamma(X,Z) \to F(X)$ so that $\delta(\gamma)(x)$ is an element of R.)

<u>By definition</u>, this function is constant on the equivalence classes defined by the equivalence relation (2.2), hence passes to the quotient to define a real-valued function on the quotient space; this function is denoted by h_δ, and has the <u>property</u> (2.6). It is convenient though to shorten all this, and say that (2.6) <u>defines</u> h_δ

Using a linear coordinate system

$$(z^a)$$

for Z, and an arbitrary coordinate system (x^i) for X, we can make this more explicit, and tie up with the more classical notation described previously.

As before, for

$$\alpha = (i_1,\ldots,i_n) \in I_+^n \quad ,$$

set:

$$\partial_\alpha = \frac{\partial^{|\alpha|}}{\partial x_1^{i_1},\ldots,\partial x_n^{i_n}} \tag{2.7}$$

Let

$$\Delta_\alpha^a: \Gamma(X,Z) \to F(Z) \tag{2.8}$$

be the linear differential operator defined by the following formula:

SYMMETRIES

$$\Delta_\alpha^a = \partial_\alpha \gamma^*(z^a) \tag{2.9}$$

Here is another way of thinking about this formula. Suppose that $\gamma \in \Gamma(X,Z)$ is defined in those coordinates by functions:

$$x \to (z^a(x)) \quad .$$

Then,

$$\Delta_\alpha^a(\gamma)(x) = \partial_\alpha z^a(x) \quad . \tag{2.10}$$

We can now denote the function h_δ as follows:

$$h_\delta = z_\alpha^a \quad .$$

We see that

$$(x, z, z_\alpha \cdots)$$

are the coordinates for $M(\Delta, X, Z)$.

3. THE LIE DIFFERENTIAL FORM SYSTEM

Keep the notation of previous sections. In particular,

$$(x^i), \quad 1 \leq i,j \leq n \quad ,$$

is a coordinate system for X. Set:

$$\partial_i = \frac{\partial}{\partial x_i}$$

For $\Delta \in D(X,Z)$, set:

$$\Delta(i) = \partial_i \Delta \quad . \tag{3.1}$$

Let Δ be an arbitrary collection of differential operators in $D(X,Z)$.

<u>Definition</u>. Δ_1 consists of the $\delta \in \Delta$ such that

$$\delta(i) \equiv \partial_i \delta \in \Delta \quad \text{for all } i \tag{3.2}$$

Δ_2 consists of the $\delta \in \Delta$ such that

$$\delta(i,j) \equiv \partial_i \partial_j \delta \in \Delta \quad \text{for all } i,j \tag{3.3}$$

and so forth.

<u>Remark</u>. If Δ consists of the r-th order differential operators, then Δ_1 consists of those of order $(r-1)$, Δ_2 of order $(r-2)$, and so forth.

<u>Definition</u>. Let $\delta \in \Delta_1$. The differential form:

$$\theta_\delta = dh_\delta - h_{\delta(i)} \, dx_i \tag{3.4}$$

is said to be the <u>contact differential form</u> associated with δ.

<u>Remark</u>. Note that $h_{\delta(i)}$, hence θ_ϕ, is well-defined on $M(\Delta,X,Z)$ only if $\delta \in \Delta_1$.

<u>Definition</u>. A <u>differential form system</u> on a manifold is a collection of differential forms such that:

    a) The sum of two forms in the system is again in the system;

SYMMETRIES

b) If ω is in the system, and ω_1 is an arbitrary form, $\omega \wedge \omega_1$ is in the system;

c) If ω is in the system, so is $d\omega$.

A <u>solution</u> of such a system is a submanifold on which each form in the system is zero.

Remark. Such an object is also called a <u>differential ideal</u> of <u>differential forms</u> or an <u>exterior differential system</u>. A "solution" is also called an <u>integral submanifold</u>. (I am trying to banish the word "integral" from the theory. It has its roots in the 18-th century meaning of solution of a differential system, which is no longer what we mean.) An arbitrary collection of differential forms <u>determines</u> a differential form system, namely the smallest set of forms containing the collection and satisfying a), b), and c).

Definition. The <u>Lie contact system</u> is the differential form system on $M(\Delta, X, Z)$ determined by the one-forms

$$\theta_\delta ,$$

where δ runs through Δ_1.

Here are some basic results about these concepts. Although we state them as "theorems", their proof is routine, and is left to the reader as exercises.

Theorem 3.1. Let $S \subset M(\Delta,X,Z)$ be a system of partial differential equations, which form a submanifold of $M(\Delta,X,Z)$. Let $\gamma: X \to Z$ be a map which is a solution of S. Then,

$$\partial\gamma: X \to S$$

is a submanifold which is a solution of the differential form system obtained by restricting the Lie differential form system to S.

Theorem 3.2. If γ is a map $: X \to Z$, then its partial derivative

$$\partial\gamma: X \to M(\Delta,X,Z)$$

is a solution of the Lie system.

Conversely, not every solution of the Lie differential form system arises from a map $X \to Z$. It was one of Lie's fundamental contributions to 19-th century differential equation theory to be the first one to clearly recognize this and to study its ramifications.

Definition. Let $S \subset M(\Delta,X,Z)$ be a differential equation system. (For simplicity, suppose from now on that it is a submanifold.) Let

$$n = \dim X .$$

Then, a __solution for__ S __in the sense of Lie__ is an n-dimensional submanifold of S which is a solution of the Lie differential form system restricted to S.

4. LIE SYMMETRIES AND BÄCKLUND SYMMETRIES OF DIFFERENTIAL EQUATION SYSTEMS

Let us recapitulate notation. X is a manifold, Z is a vector space. Δ is a collection of linear differential operators

$$\Delta: \Gamma(X,Z) \to F(X)$$

$M(\Delta,X,Z)$ is the mapping element space defined by Δ. Denote now by

$$\underset{\sim}{L}$$

the Lie differential form system on $M(\Delta,X,Z)$, determined by the contact-form (3.4). (In GPS I called this the <u>contact differential form system</u>.)

Let $S \subset M(\Delta,X,Z)$ be a subset which determines a <u>differential equation system</u>, with <u>independent variables</u> X, <u>dependent variables</u> Z. For simplicity, we suppose S is a <u>submanifold</u>. Let

$$\underset{\sim}{L}_S = \text{restriction of } \underset{\sim}{L} \text{ to } S \;.$$

<u>Definition</u>. A <u>Lie symmetry</u> of S is a map

$$\phi: M(\Delta,X,Z) \to M(\Delta,X,Z)$$

satisfying the following conditions:

$$\phi(S) \subset S \tag{4.1}$$

$$\phi^*(\underset{\sim}{L}) \subset \underset{\sim}{L} \tag{4.2}$$

Remark. In particular, φ maps a "solution of S in the sense of Lie" into another one.

One might also consider a weaker notion of "symmetry", namely:

A map φ: S → S such that

$$\phi^*(\underset{\sim}{L}_S) \subset \underset{\sim}{L}_S$$

Such a map might be called a <u>symmetry in the sense of Cartan</u>, but, at least at the moment, I will forego their consideration.

Thus, Lie symmetries give "geometric" ways of mapping the set of solutions of S into other solutions. The "Bäcklund transformation" is more subtle, and less clear as to its geometr and group-theoretic meaning. Unfortunately, the classical autho have not left us a clear and/or precise definition of what exact is meant by a Bäcklund transformation. I will adopt for my form alization the informal description given in the Introduction to Cartan's paper "Sur les transformations de Bäcklund", in Part II Volume 2 of his Collected Works. It should then be clear to the reader that this "definition" is not rigid, and might profitably be altered.

To define a Bäcklund transformation, one should be given another set

$$\Delta'$$

of linear differential operators : $\Gamma(X,Z) \to F(X)$.

Remark. In the typical application,

Δ = set of second order differential operators

Δ' = set of first order differential operators

Suppose also given a subset

$B \subset M(\Delta,X,Z)$.

pair (γ_1,γ_2) of maps : $X \to Z$ are said to be Bäcklund trans-
orms of each other with respect to B, or to be a Bäcklund
air, if the following conditions are satisfied:

$(\partial\gamma_1(x),\partial\gamma_2(x)) \in B$

for all $x \in X$.

finition. Let $S \subset M(\Delta,X,Z)$ be a differential equation system.
Bäcklund transformation $B \subset M(\Delta',X,Z)$ is said to be a Bäcklund
mmetry of S if there is at least one Bäcklund pair (γ_1,γ_2)
$\Gamma(X,Z) \times \Gamma(X,Z)$ such that:

$\gamma_1 \neq \gamma_2$

γ_1 and γ_2 are solutions of S.

To see what this means, it is best to see an example. The
earest one seems to be provided in Darboux' "Theorie des Surfaces"
rt III, Book VII, Chapter XII.

In this example

$X = R^2$, $Z = R$.

Δ = set of second order differential operators

Δ' = set of first order differential operators.

Denote coordinates on X by

$$(x,y) ,$$

coordinates on Z by z. The Lie coordinates on

$$M(\Delta,X,Z)$$

may then be denoted by

$$(x,y,z,z_x,z_y,z_{xx},z_{yy},z_{xy}) .$$

Lie coordinates on $M(\Delta',X,Z)$ are:

$$(x,y,z,z_x,z_y) .$$

The system S is defined by a single equation:

$$z_{xx} - z_{yy} = \sin z \cos z . \tag{4}$$

Consider another copy of $M(\Delta',X,Z)$, with Lie coordinates on it denoted by

$$(x',y',z',z'_x,z'_y) .$$

Regard $M(\Delta',X,Z) \times M(\Delta',X,Z)$ as defined by the following coordinate system:

$$(x,y,z,z_x,z_y,x',y',z',z'_x,z'_y) .$$

Let B be the subset of

SYMMETRIES

$$M(\Delta',X,Z') \times M(\Delta',X,Z)$$

defined by the following equations:

$$\begin{aligned} x &= x' \\ y &= y' \\ z_x + z'_x &= a \sin(z-z') \\ z_y - z'_y &= b \sin(z+z') \end{aligned} \quad (4.3)$$

(a,b are real constants.) B defines the <u>Bäcklund transformation</u>. A pair of functions

$$(\{z(x,y)\}, \{z'(x,y)\}) \ ,$$

i.e., maps $\gamma,\gamma': X \to Z$, is a <u>Bäcklund pair</u> if they satisfy the following system of partial differential equations:

$$\begin{aligned} \partial_x z + \partial_x z' &= a \sin(z-z') \\ \partial_y z - \partial_y z' &= b \sin(z+z') \end{aligned} \quad (4.4)$$

We want to find differential equation systems S with the property that pairs of solutions are Bäcklund related via B. Darboux' technique for this is to fix z', then to ask what conditions it must satisfy in order that the <u>integrability conditions</u> are satisfied for the resulting first order system for z. To do this, apply ∂_y to the first equation of (4.4), and ∂_x to the second, which is the usual method for finding integrability conditions:

$$\partial_{yx}z = -\partial_{yx}z' + a\cos(z-z')(\partial_y z - \partial_y z')$$

$$= -\partial_{yx}z' + a\cos(z-z')(\partial_y z' + b\sin(z+z')) - a\cos(z-z')\partial_y z'$$

$$= -\partial_{yx}z' + ab\cos(z-z')\sin(z+z') \qquad (4.4)$$

$$\partial_{xy}z = \partial_{xy}z' + b\cos(z+z')(\partial_x z + \partial_x z')$$

$$= \partial_{xy}z' + b\cos(z+z')(a\sin(z-z') - \partial_x z') + b\cos(z+z')\partial_x z$$

$$= \partial_{xy}z' + ab\cos(z+z')\sin(z-z') \qquad (4.5)$$

To derive the integrability conditions, equate the right hand side of (4.4) and (4.5);

$$2\partial_{xy}z' = ab(\cos(z-z')\sin(z+z') - \cos(z+z')\sin(z-z'))$$

$$= ab\sin(2z') \qquad (4.6)$$

This is a second order differential equation which must be satisfied if a Bäcklund pair <u>is to exist</u>. Here are the basic results:

SYMMETRIES

Theorem 4.1. Define B as a subset of $M(\Delta',X,Z) \times M(\Delta',X,Z)$, via equations (4.3). Then, given the map $\gamma': X \to Z$, defined by a function $z'(x,y)$, there is a map γ such that

$$(\gamma,\gamma')$$

is a <u>Bäcklund pair</u> if and only if γ' satisfies the second order differential equation (4.6). (It has recently been called the <u>Sine-Gordon equation</u>, and has played an important role in work on non-linear wave propagation.)

<u>Exercise</u>. If (γ,γ') are a Bäcklund pair, associated with the subspace B, show that γ <u>also</u> satisfies the Sine-Gordon equation (4.6).

Given γ', a solution of (4.6), we can <u>solve</u> (4.4) as a differential equation for γ. However, the solution is not <u>unique</u>. Thus, the Bäcklund "transformation" is not really a "mapping" of the space of solutions of (4.6) into itself, but merely a set of <u>ordered pairs</u>. This is probably why modern mathematics has never considered these objects, until their recent appearance in Applied Mathematics. (This is a good illustration that Applications are necessary for even the <u>internal</u> development of Pure Mathematics.)

The Sine-Gordon equation (4.6) appeared in the 19-th century as the differential equation which determines the isometric embedding of two-dimensional Riemannian manifolds of constant curvature. This is the context in which it is discussed by Darboux. One might expect that there are other relations of this type between Applied Mathematics and Geometry!

Of course, at the "mapping element" level a "Bäcklund transformation" is really a transformation. Consider $M(\Delta',X,Z)$ as a fiber space over X, i.e.,

$$(x,y,z,z_x,z_y) \to (x,y) \ .$$

Then, formulas (4.3) define a map

$$B: M(\Delta',X,Z) \to M(\Delta',X,Z)$$

which is "fiber-preserving", i.e., leads to a commutative diagram of maps:

$$\begin{array}{ccc} M(\Delta',X,Z) & \to & M(\Delta',X,Z) \\ \downarrow & & \downarrow \\ X & = & X \end{array}$$

Chapter VII

THE PROLONGATION STRUCTURE OF ESTABROOK AND WAHLQUIST, THE DIFFERENTIAL GEOMETRY OF SOLITONS, AND CARTAN-EHRESMANN CONNECTIONS

1. INTRODUCTION

Recent work on non-linear partial differential equations has introduced many new ideas to applied mathematics. A variety of concepts--the inverse scattering techniques, Bäcklund transformations, conservation laws, "soliton" solutions--have been developed to apply to a certain type of non-linear partial differential equations. So far, the equations that can be discussed only involve two independent variables. (Physically, they are "time", and one space variable.) If the concepts could be generalized to partial differential equations which depend on more than two variables, there might be available enormously important and useful mathematical tools to handle a wide variety of physical and engineering phenomena. In particular, the mathematical nature of elementary particles might be elucidated for the first time. (Perhaps our current ideas--based as they are on "perturbation" of linear models--are inadequate to deal with the observed phenomena.)

Unfortunately, all of these potential benefits are restricted by the relatively narrow framework into which the ideas have been fitted up to now. I believe that <u>differential geometry</u> might provide techniques for the extension and elaboration. One piece of evidence for this belief is a paper "Prolongation structures of non linear evolution equations", by H. Wahlquist and F. Estabrook in the Journal of Mathematical Physics, Vol. 16 1975, pp. 1-7. Indeed, this might turn out to be one of the most important papers to have appeared in mathematical physics in recent times. In this chapter I will discuss some differenti geometric ideas that are implicit or explicit in this paper, particularly in relation to Cartan's theory of Exterior Differential Systems.

I would like to thank Frank Estabrook and Hugo Wahlquist for their patient efforts to teach me their techniques and insights.

2. THE BASIC NOTION OF "PROLONGATION"

I will now attempt to describe their basic idea in as simple and clear a manner as possible using differential form language. Let M be a manifold. Consider two sorts of objects on M:

I   a <u>differential ideal</u> of differential forms on M

P   a <u>Pfaffian system</u>, i.e., an $F(M)$-submodule of the set of differential one-forms on M.

PROLONGATION STRUCTURE

$F^*(M)$ denotes the exterior algebra of differential forms on M.

<u>Definition</u>. P is a <u>prolongation of</u> I if the following condition is satisfied:

$$dP \subset F^*(M) \wedge P + I . \qquad (2.1)$$

This relation is a generalization of more familiar concepts. First, if $I = 0$, (2.1) expresses the fact that P is <u>completely integrable</u>. The (local) Frobenius complete integrability theorem then asserts (in one formulation) that there are (locally) one forms $\omega_1,\ldots,\omega_n \in P$ which form an F(M)-basis for P and such that

$$d\omega_1 = 0 \cdots = d\omega_n .$$

Second, if P is generated by a single element ω such that

$$d\omega \in I ,$$

then ω is a <u>conservation law</u> for I. (See IM, Vol. 5.)

I believe that studying relation (2.1) (and its possible generalizations to higher degree forms) will be an important topic for the study of the geometric properties of solutions of non-linear partial differential equations.

In their paper, Estabrook and Wahlquist mainly start off with I, then look for P. This is, of course, important in the applications where I is given by "nature" as a system of partial differential equations. I believe that it might be equally useful (at least for understanding the mathematical nature of the material) to start off with P, and find I. This procedure will be the main topic of this chapter.

3. A PROLONGATION OF THE KORTEWEG-DE VRIES EQUATION

In their basic paper [1] Wahlquist and Estabrook have described various prolongations of the Korteweg-de Vries equation. In this section I will isolate one particular calculation, which illustrates well their ideas.

Consider R^6, with variables labelled as:

$$(x,t,u,p,z,y) \ .$$

Set:

$$\omega = dy + \omega_1 + y\omega_2 + y^2\omega_3 \tag{3.1}$$

where

$$\omega_1 = (2u-\lambda)dx - 4[(u+\lambda)(2u-\lambda)+\tfrac{1}{2}p] \, dt \tag{3.2}$$

$$\omega_2 = 4z \, dt \tag{3.3}$$

$$\omega_3 = dx - 4(u+\lambda) \, dt \tag{3.4}$$

PROLONGATION STRUCTURE

Notice that $\omega_1, \omega_2, \omega_3$ <u>do not involve the variable</u> y. λ is a parameter, that is considered to be a constant. Set:

$$\alpha_1 = du \wedge dt - z\, dx \wedge dt$$

$$\alpha_2 = dz \wedge dt - p\, dx \wedge dt$$

$$\alpha_3 = -du \wedge dx + 12uz\, dx \wedge dt + dp \wedge dt$$

Then, they show that:

$$\boxed{\begin{aligned} d\alpha_1 &= dx \wedge \alpha_2 \\ d\alpha_2 &= dx \wedge \alpha_3 \\ d\alpha_3 &= -12 dx \wedge (z\alpha_1 + u\alpha_2) \end{aligned}} \qquad (3.5)$$

Let I be the differential ideal of forms generated by $\alpha_1, \alpha_2, \alpha_3$. Relations (3.5) show that I is generated <u>as an exterior algebra ideal</u> by $\alpha_1, \alpha_2, \alpha_3$ alone.

I defines the <u>Korteweg-de Vries equation</u>. To see this, consider a two dimensional solution manifold of I <u>on which</u> dx <u>and</u> dt <u>are linearly independent</u>. Such a submanifold can be written locally as

$$(x,t) \to (u(t,x), p(t,x), z(t,x))$$

Estabrook and Wahlquist show that the "reduction manifold"

condition for I is equivalent to the following relations:

$$z = \frac{\partial u}{\partial x}, \quad p = \frac{\partial z}{\partial x} = \frac{\partial^2 u}{\partial x^2}$$
$$\frac{\partial u}{\partial t} + \frac{\partial p}{\partial x} + 12uz = 0 \tag{3.6}$$

These relations imply that $(t,x) \to u(t,x)$ satisfy the Korteweg-de Vries equation:

$$\frac{\partial u}{\partial t} + \frac{\partial^2 u}{\partial x^3} + 12u \frac{\partial u}{\partial x} = 0 \tag{3.7}$$

Formula (4.6) of their paper is of the most interest to us:

$$d\omega = -4(4u + y^2 + \lambda)\alpha_1 + 4y\alpha_2 - 2\alpha_3$$
$$+ \text{(an element of } F^*(M) \wedge P) \tag{3.8}$$

It is relation (2.1), in this case. Wahlquist and Estabrook go on to show how the Bäcklund transformations, soliton solutions, conservation laws, etc. may be obtained from this relation, and others closely related to it.

Here is more terminology that will be useful. I do not want to fix it with a formal definition, hence will describe it only in an informal way. Suppose M has a set of coordinates labelled

$$(y, x^i) \;.$$

Suppose that I has a basis consisting of differential forms in the x <u>alone</u>. The y's will then be called (following Estabrook and Wahlquist) <u>pseudopotentials</u> for the system of differential equations defined by I. The form ω in P might by analogy called <u>pseudo-conservation laws</u> for the system I. Now I turn to a description of some general ideas which underly the examples treated so brilliantly by Wahlquist and Estabrook.

4. SYSTEMS WHICH ADMIT A PSEUDOCONSERVATION LAW WHICH IS QUADRATIC IN THE PSEUDOPOTENTIAL

We can now abstract from Wahlquist and Estabrook's work a simple general situation which includes many of the equations which admit solitions (Korteweg-de Vries, Sine-Gordon, non-linear Schrödinger,...). Let X and Y be manifolds, and let

$$M = Y \times X \;.$$

For the moment, suppose that Y is one-dimensional, with its variable denoted by y. Suppose also that I is an ideal of differential forms on the space X, whose solution submanifolds defines a system of partial differential equations.

Suppose given a one-form ω on M such that:

$d\omega$ lies in the exterior ideal generated by I and ω (4.1)

Condition (4.1) defines ω as a <u>pseudoconservation law for the system</u> I.

<u>Definition</u>. ω is said to be a <u>quadratic pseudoconservation law</u> (with respect to the variable y) if it is of the following form:

$$\omega = \omega_0 + y\omega_1 + y^2\omega_2 - dy \qquad (4.2)$$

where $\omega_0, \omega_1, \omega_2$ are one-forms on X.

Let us now work out the consequences of (4.1) and (4.2). Apply d to both sides of (4.2):

$$\begin{aligned}
d\omega &= d\omega_0 + dy \wedge \omega_1 + y d\omega_1 + 2y\, dy \wedge \omega_2 + y^2 d\omega_2 \\
&= d\omega_0 + (\omega_0 + y\omega_1 + y^2\omega_2 - \omega) \wedge \omega_1 + y d\omega_1 \\
&\quad + 2y(\omega_0 + y\omega_1 + y^2\omega_2 - \omega) \wedge \omega_2 + y^2 d\omega_2 \\
&= d\omega_0 - \omega_0 \wedge \omega_1 + y(d\omega_1 + 2\omega_0 \wedge \omega_2) + y^2(d\omega_2 + \omega_2 \wedge \omega_1 + 2\omega_1 \wedge \omega_2) \\
&\quad + (\text{terms in } \omega)
\end{aligned}$$

PROLONGATION STRUCTURE

This formula provides us with the following

Theorem 4.1. The exterior differential system I admits the form ω given by (4.2) as a pseudoconservation law if and only if the following relations are satisfied:

$$
\begin{array}{|l|}
\hline
d\omega_0 - \omega_0 \wedge \omega_1 \in I \\
\\
d\omega_1 + 2\omega_0 \wedge \omega_2 \in I \\
\\
d\omega_2 + \omega_1 \wedge \omega_2 \in I \\
\hline
\end{array}
\tag{4.3}
$$

What is the meaning of these relations? Obviously, there is both a geometric and analytical significance. The first involves the theory of <u>Cartan-Ehresmann connections</u> (see Vol. 10 of IM), the theory of <u>Lie algebra valued differential forms</u>, and the <u>theory of curvature</u>. I will briefly go into these topics later on in this chapter. Analytically, recall that the <u>solutions</u> of the underlying differential equations are solution submanifolds of I, i.e., submanifold maps

$$\phi: N \to X$$

such that

$$\boxed{\phi^*(I) = 0} \tag{4.4}$$

Combine (4.3) and (4.4):

$$
\begin{aligned}
d\phi^*(\omega_0) &= \phi^*(\omega_0) \wedge \phi^*(\omega_1) \\
d\phi^*(\omega_1) &= -2\phi^*(\omega_0) \wedge \phi^*(\omega_2) \\
d\phi^*(\omega_2) &= -\phi^*(\omega_1) \wedge \phi^*(\omega_2)
\end{aligned}
\qquad (4.5)
$$

These equations have an important Lie-group theoretical significance, which we now explain.

Exercise. Let G be the Lie group $SL(2,R)$, the group of 2×2 real matrices of determinant one. Show that a basis $(\theta_1, \theta_2, \theta_3)$ of left-invariant one-forms on G can be chosen so that:

$$
\begin{aligned}
d\theta_0 &= \theta_0 \wedge \theta_1 \\
d\theta_1 &= -2\theta_0 \wedge \theta_2 \\
d\theta_2 &= -\theta_1 \wedge \theta_2
\end{aligned}
\qquad (4.
$$

Here is the significance of these relations. If N is simply connected, there is a map

PROLONGATION STRUCTURE

such that:
$$\gamma_\phi : N \to G$$

$$\boxed{\begin{aligned} \gamma^*(\theta_i) &= \phi^*(\omega_i) \\ \text{for } 0 &\leq i \leq 2 \end{aligned}} \quad (4.7)$$

Thus, to each solution ϕ of the differential equations defining I (e.g., Korteweg-de Vries, Sine-Gordon, non-linear Schrödinger, etc.) there is assigned--by relation (4.7)--a map $\gamma_\phi : N \to G$, i.e., a submanifold of G. Understanding the geometric significance of this assignment $\phi \to \gamma_\phi$ is obviously at the heart of understanding the geometric meaning of the concepts introduced by Estabrook and Wahlquist.

There are two ways of understanding these relations in terms of modern differential geometry--in terms of the theory of Lie algebra valued differential forms, and the theory of Cartan-Ehresmann connections. We now deal with the first, since it is obviously a direct generalization of the formulas given above, particularly (4.3). In fact, we shall show that the left side of (4.3) involves a differential operator which is very naturally defined in terms of Lie algebra-valued differential forms. In turn, the theory of these objects will be related to the theory of Cartan-Ehresmann connections. (See Vol. 10 of IM.)

5. LIE ALGEBRA VALUED DIFFERENTIAL FORMS

Let X be a manifold. The reader should be used to the definition of a differential form on X. There are several alternate definitions. The most direct one is to say that an r-th degree form is a mapping

$$\omega: V(X) \times \ldots \times V(X) \to F(X)$$

which is $F(X)$-multilinear and skew-symmetric.

Various generalizations of this concept are useful in differential geometry. The one concerning us is the notion of "Lie algebra valued differential form". Let $\underset{\sim}{G}$ be a Lie algebra with the real numbers as field of scalars. An r-<u>th degree</u>, $\underset{\sim}{G}$-<u>valued differential form</u> is a mapping

$$\underset{\sim}{\omega}: V(X) \times \ldots \times V(X) \to M(X, \underset{\sim}{G})$$

which is $F(X)$-multilinear and skew-symmetric. ($M(X,G)$ denotes the space of mappings $X \to \underset{\sim}{G}$.)

The usual operations of differential geometry, e.g., Lie derivative, exterior derivative and contraction, can be defined on these objects, in a direct generalization of the classical case. Exterior product is not defined as usual: Instead, it is replaced by an operation using the given Lie algebra structure on $\underset{\sim}{G}$. We denote it by

$$(\underset{\sim}{\omega}_1, \underset{\sim}{\omega}_2) \to [\underset{\sim}{\omega}_1, \underset{\sim}{\omega}_2] \ .$$

PROLONGATION STRUCTURE

As definition, suppose that $\underset{\sim}{\omega}_1$ is an r-form, while $\underset{\sim}{\omega}_2$ is an s-form. $[\underset{\sim}{\omega}_1,\underset{\sim}{\omega}_2]$ is to be an (r+s)-form. For $A_1,\ldots,A_{r+s} \in V(X)$,

$$[\underset{\sim}{\omega}_1,\underset{\sim}{\omega}_2](A_1,\ldots,A_{r+s}) = \sum \pm [\underset{\sim}{\omega}_1(A_{i_1},\ldots,A_{i_s}),\underset{\sim}{\omega}_2(A_{i_{r+1}},\ldots,A_{i_{r+s}})]$$

(5.1)

The sum on the right hand side of (5.1) is over all permutations of $(1,\ldots,r+s)$. $\pm$ denotes the signature of the permutation.

We shall actually only need this formula for $r = s = 1$. In this case,

$$[\underset{\sim}{\omega}_1,\underset{\sim}{\omega}_2](A_1,A_2) = [\underset{\sim}{\omega}_1(A_1),\underset{\sim}{\omega}_2(A_2)] - [\underset{\sim}{\omega}_1(A_2),\underset{\sim}{\omega}_2(A_1)]$$

(5.2)

In particular, notice that the skew-symmetry of the Lie algebra operation forces the following formula:

$$[\underset{\sim}{\omega},\underset{\sim}{\omega}](A_1,A_2) = 2[\underset{\sim}{\omega}(A_1),\underset{\sim}{\omega}(A_2)]$$

for $A_1,A_2 \in V(X)$.

(5.3)

Lie algebra-valued differential forms can be described in terms of scalar-valued differential forms by introducing a basis for $\underset{\sim}{G}$. Suppose:

$$\dim \underset{\sim}{G} = m .$$

Introduce indices and the summation convention:

$$1 \leq a,b \leq m .$$

Let

$$(W_a)$$

be a basis for $\underset{\sim}{G}$. The **structure constants**

$$(C^c_{ab})$$

of $\underset{\sim}{G}$ with respect to this basis are defined as follows:

$$\boxed{[W_a, W_b] = C^c_{ab} W_c} \tag{5}$$

A $\underset{\sim}{G}$-valued form on X can then be written as

$$\underset{\sim}{\omega} = \omega^a W_a , \tag{5}$$

where (ω^a) are scalar valued forms, called its **components**.

For example, here is the formula for (5.3) in terms of components:

$$[\underset{\sim}{\omega}, \underset{\sim}{\omega}](A_1, A_2) = 2[\omega^a(A_1)W_a, \omega^b(A_2)W_b]$$

$$= 2\omega^a(A_1)\omega^b(A_2)[W_a, W_b]$$

$$= 2\omega^a(A_1)\omega^b(A_2)C^c_{ab}W_c$$

$$= C^c_{ab}(\omega^a \wedge \omega^b)(A_1, A_2)$$

PROLONGATION STRUCTURE

Hence,

$$[\underset{\sim}{\omega}, \underset{\sim}{\omega}] = (C^c_{ab}\omega^a \wedge \omega^b) W_c \tag{5.6}$$

Let G be a Lie group whose Lie algebra is $\underset{\sim}{G}$. A special sort of $\underset{\sim}{G}$-valued form can be defined on G. It is called the <u>Maurer-Cartan form</u>, defined as follows

$$\underset{\sim}{\theta}(A) = \text{the element of } \underset{\sim}{G} \text{ which has the same value as } A \text{ at the identity element of } G. \tag{5.7}$$

($\underset{\sim}{G}$ is identified with the <u>left-invariant</u> vector fields on G.)

<u>Exercise</u>. If (W_a) is a basis for $\underset{\sim}{G}$, and (θ^a) is the dual basis for left-invariant one-forms, i.e.,

$$\theta^a(W_b) = \delta^a_b ,$$

show that:

$$\underset{\sim}{\theta} = \theta^a W_a \tag{5.8}$$

Using (5.8), show that:

PROLONGATION STRUCTURE

$$\boxed{d\underset{\sim}{\theta} = -\frac{1}{2}[\underset{\sim}{\theta},\underset{\sim}{\theta}]} \tag{5.9}$$

This is called the <u>Maurer-Cartan equation</u>.

<u>Theorem 5.1</u>. Let $\underset{\sim}{\omega}$ be a G-valued one-form on a manifold X, which satisfies the equation:

$$d\underset{\sim}{\omega} = -\frac{1}{2}[\underset{\sim}{\omega},\underset{\sim}{\omega}] \quad . \tag{5.1}$$

Suppose also that X is <u>simply connected</u>. Then, there is a ma

$$\phi: X \to G$$

such that:

$$\boxed{\phi^*(\underset{\sim}{\theta}) = \underset{\sim}{\omega}} \tag{5.1}$$

where $\underset{\sim}{\theta}$ is the Maurer-Cartan form on G.

The proof is left as an <u>exercise</u>. The idea is to constru the manifold

$$X \times G \quad ,$$

and the one-forms

$$\underset{\sim}{\omega} - \underset{\sim}{\theta}$$

It is seen that relations (5.10) are the <u>integrability conditi</u> for the Pfaffian system defined by these forms. The map we are looking for is obtained by identifying its graph with a maximal solution submanifold of this Pfaffian system.

Definition. Let $\underset{\sim}{\omega}$ be a Lie algebra valued one-form on a manifold X. The two-forms

$$\underset{\sim}{\Omega} = d\underset{\sim}{\omega} + \frac{1}{2}\,[\underset{\sim}{\omega},\underset{\sim}{\omega}] \tag{5.12}$$

is called the curvature of $\underset{\sim}{\omega}$.

Example: The Maxwell-Yang-Mills field.

See "Fourier analysis...", "Vector bundles..." and IM, Vol. 10. X is R^4, the "space-time" of special relativity. (x^μ), $0 \leq \mu \leq 3$, are Cartesian coordinates, i.e., $x^0 = t$ is "time", $\vec{x} = (x^1, x^2, x^3)$ are "space" coordinates. $\underset{\sim}{G}$ is an arbitrary Lie algebra. A Maxwell-Yang-Mills field is a pair $(\underset{\sim}{\omega}, \underset{\sim}{\Omega})$, where

$\underset{\sim}{\omega}$ is a $\underset{\sim}{G}$-valued one-form on X

$\underset{\sim}{\Omega}$ is a $\underset{\sim}{G}$-valued two-form on X.

Introduce its components:

$$\begin{aligned} \underset{\sim}{\omega} &= \omega^a W_a \\ \underset{\sim}{\Omega} &= \Omega^a W_a \end{aligned} \tag{5.13}$$

$$\boxed{\begin{aligned} \omega^a &= A_\mu^a \, dx^\mu \\ \Omega^a &= F_{\mu\nu}^a \, dx^\mu \wedge dx^\nu \end{aligned}}$$

Of course, in the physics literature one usually sees $\underset{\sim}{\omega}$ and $\underset{\sim}{\Omega}$ defined in terms of its components (A_μ^a), $(F_{\mu\nu}^a)$. For example, "Maxwell" is obtained by taking: $\dim \underset{\sim}{G} = 1$.

We can now write the field equations as follows:

$$\underset{\sim}{\Omega} = d\underset{\sim}{\omega} + \frac{1}{2}[\underset{\sim}{\omega},\underset{\sim}{\omega}] \, , \tag{5.14}$$

$$\delta\underset{\sim}{\Omega} = 0 \, , \tag{5.15}$$

where δ is the "Hodge" operator suitably generalized by means of a symmetric bilinear form on $\underset{\sim}{G}$. (If $\underset{\sim}{G}$ is semisimple, the Killing form may be used, to give the most "covariant" form of the equations.)

The key to "gauge invariance" is (5.14), which expresses Ω as the "curvature". (This was presumably the key fact in Yang and Mills original work--a generalization of Einstein's gravitational equation.) In fact, consider

$M(X,G)$,

the _group_ (under pointwise multiplication) of maps $X \to G$. (In my previous work, I call this the _gauge group_.)

PROLONGATION STRUCTURE

Exercise. Suppose $\alpha \in M(X,G)$. Define

$$\boxed{\alpha(\underset{\sim}{\omega}) = \alpha \omega \alpha^{-1} - \alpha^{-1} d\alpha} \tag{5.16}$$

$$\boxed{\alpha(\underset{\sim}{\Omega}) = \alpha \Omega \alpha^{-1}} \tag{5.19}$$

Show that α <u>commutes</u> with the differential operator (5.14). It does not commute with δ, but does map solutions of $\delta = 0$ into solutions. Thus, $M(X,G)$ is a "symmetry group" of the Maxwell-Yang-Mills fields.

6. LIE ALGEBRA VALUED ONE-FORMS AND CARTAN-EHRESMANN CONNECTIONS

I will continue to describe various general facts, in a form that will be useful to us in our study of "pseudoconservation laws" of partial differential equations. Let X continue to be an arbitrary manifold. $\underset{\sim}{G}$ is a Lie algebra. Y is another manifold.

Suppose that $\underset{\sim}{G}$ acts <u>as a Lie algebra of vector fields on</u> Y. Construct the product space

$$X \times Y .$$

Let

$$\pi: X \times Y \to X$$

be the Cartesian projection, i.e.,

$$\pi(x,y) = x \quad.$$

makes $X \times Y$ into a __fiber space__ over X.

Suppose that $\underset{\sim}{\theta}$ is a $\underset{\sim}{G}$-valued one-differential form on X. Recall (e.g., from Vol. X) the concept of __connections__ for such a fiber space. It can either be defined by a set of vector fields on $X \times Y$ --called the horizontal vector fields-- or by a dual set of one-forms. We shall work with forms, and show that $\underset{\sim}{\theta}$ defines such a connection with π.

Let (W_a) be a basis for $\underset{\sim}{G}$, as before. Let

$$(y^\alpha) \quad, \qquad 1 \le \alpha, \beta \le p \quad,$$

be a set of coordinates for Y. The W_a are __given__ as vector fields on Y. Suppose that, in these coordinates, they are given by the following formulas:

$$W_a = W_a^\alpha \frac{\partial}{\partial y^\alpha} \quad, \tag{6.}$$

where $(W_a^\alpha(y))$ are functions of y. Suppose that $\underset{\sim}{\theta}$ is give by:

$$\underset{\sim}{\theta} = \theta^a W_a \quad. \tag{6.}$$

Now, define:

$$\boxed{\omega^\alpha = dy^\alpha + W^\alpha_a \theta^a} \tag{6.3}$$

These one-forms define the connection, i.e., the vector fields A on $X \times Y$ such that

$$\omega^\alpha(A) = 0$$

are defined to be the <u>horizontal</u> vector fields.

Here is a more coordinate-free way to define the connection. We, of course, consider $\partial/\partial y^\alpha$ as vector fields on $X \times Y$. They are <u>vertical</u>, i.e., tangent to the fibers of π. Analytically,

$$\pi_* \left(\frac{\partial}{\partial y^\alpha} \right) = 0 \ .$$

Now set:

$$\boxed{\underset{\sim}{\omega} = \omega^\alpha \otimes \frac{\partial}{\partial y^\alpha}} \tag{6.4}$$

$\underset{\sim}{\omega}$ is a linear map from the tangent bundle of $X \times Y$ to the vector bundle on $X \times Y$ consisting of vertical vectors. $\underset{\sim}{\omega}$ may be called a <u>vector-bundle valued differential form</u>. (Thus, a vector-valued differential form is one taking values in a product vector bundle.) The tangent vectors such that $\omega(v) = 0$ are called the <u>horizontal vectors</u>.

Exercise. Show that $\underset{\sim}{\omega}$ is independent of the coordinate system (y) and the Lie algebra basis (W_a) used to define it.

Combining (6.3) and (6.4) gives another useful formula:

$$\underset{\sim}{\omega} = dy^\alpha \otimes \frac{\partial}{\partial y^\alpha} + W_a^\alpha \theta^a \otimes \frac{\partial}{\partial y^\alpha}$$

$$\boxed{= dy^\alpha \otimes \frac{\partial}{\partial y^\alpha} + \theta^a \otimes W_a} \qquad (6.5)$$

We can also write this as

$$\boxed{\underset{\sim}{\omega} = dy^\alpha \otimes \frac{\partial}{\partial y^\alpha} + \underset{\sim}{\theta}} \qquad (6.6)$$

This formula exhibits most clearly the role that the Lie algebra valued form $\underset{\sim}{\theta}$ plays in the definition of the connection.

We can now motivate the <u>curvature</u> of the Lie algebra valued form as the geometric object which decides the <u>complete integrability</u> of the connection. To this end, let us apply the exterior derivative operation to both sides of (6.3). Set:

$$\partial_\alpha = \frac{\partial}{\partial y^\alpha} \quad .$$

PROLONGATION STRUCTURE

Then,

$$
\begin{aligned}
d\omega^\alpha &= \partial_\beta(W_a^\alpha)dy^\beta \wedge \theta^a + W_a^\alpha d\theta^a \\
&= \partial_\beta(W_a^\alpha)W_b^\beta \theta^b \wedge \theta^a + W_a^\alpha d\theta^a + \cdots
\end{aligned}
\qquad (6.7)
$$

(The terms ... indicate terms involving the ω^α.)

Now,

$$[W_a, W_b] = C_{ab}^c W_c \quad,$$

hence, using (6.1),

$$
\begin{aligned}
[W_a^\alpha \partial_\alpha, W_b^\beta \partial_\beta] &= W_a^\alpha \partial_\alpha(W_b^\beta)\partial_\beta - W_b^\beta \partial_\beta(W_a^\alpha)\partial_\alpha \\
&= C_{ab}^c W_c^\beta \partial_\beta \quad,
\end{aligned}
$$

or:

$$
\boxed{C_{ab}^c W_c^\beta = W_a^\alpha \partial_\alpha(W_b^\beta) - W_b^\alpha \partial_\alpha(W_a^\beta)}
\qquad (6.8)
$$

Hence,

$$
\begin{aligned}
\partial_\beta(W_a^\alpha)W_b^\beta \theta^b \wedge \theta^a &= \tfrac{1}{2}(\partial_\beta(W_a^\alpha)W_b^\beta - \partial_\beta(W_a^\alpha)W_a^\beta)\theta^b \wedge \theta^a \\
&= \text{using (6.8)}, \\
&\quad \tfrac{1}{2} C_{ab}^c W_c^\alpha \theta^a \wedge \theta^b
\end{aligned}
$$

Substitute this formula into (6.7):

$$d\omega^\alpha = \frac{1}{2} C^c_{ab} W^\alpha_c \theta^a \wedge \theta^b + W^\alpha_c d\theta^c + \cdots$$

$$= W^\alpha_c (\frac{1}{2} C^c_{ab} \theta^b \wedge \theta^a - d\theta^c) + \cdots \qquad (6.9)$$

Set:

$$\Omega^c = d\theta^c + \frac{1}{2} C^c_{ab} \theta^a \wedge \theta^b \qquad (6.10)$$

Ω^c are called the <u>curvature forms</u>. In terms of the Lie <u>algebra</u> <u>valued</u> curvature forms $\underset{\sim}{\Omega}$ defined by (5.12), we have:

$$\underset{\sim}{\Omega} = \Omega^c W_c \qquad (6.11)$$

Here is the main geometric property:

<u>Theorem 6.1</u>. The connection is <u>flat</u>, i.e., the Pfaffian system

$$\omega^\alpha = 0$$

is completely integrable, if and only if the curvature form Ω^a vanishes.

<u>Definition</u>. Connections defined in this way by $\underset{\sim}{G}$-valued differential forms will be called <u>Cartan-Ehresmann connections with structure group</u> G. We also call them, for short, G-<u>connection</u>

<u>Remark</u>. Of course, this "definition" has the defect that it is local, dependent on the way the fiber space $\pi: E \to X$ is exhibited on a local product $X \times Y$. A definition that avoided

PROLONGATION STRUCTURE   189

this "defect" (although the one we use is perfectly adequate--
and indeed even optimal in most cases--for most applications)
involves the full artillery of the theory of fiber bundles.
See Kobayashi and Nomizu's "Foundations of differential geometry"
for the full story, at least as far as it has been written down
in the literature. (Unfortunately, it is mostly what mathematicians call "generalized nonsense"--and rather tedious in detail.)
   Notice another important property of these connections:

> Their holonomy groups--considered as transformation groups on the fibers, i.e., Y--
> are subgroups of G.

7.   THE BIANCHI IDENTITY. THE FROBENIUS CONDITION FOR
     TWO-FORMS

   Return to the situation considered in Section 5. X is a
manifold, $\tilde{G}$ is a Lie algebra, $\tilde{\omega}$ is a G-valued one-form
on $\tilde{G}$,

$$\tilde{\Omega} = d\tilde{\omega} + \frac{1}{2}[\tilde{\omega},\tilde{\omega}] \qquad (7.1)$$

is its <u>curvature form</u>. The relations resulting from applying
the exterior derivative d to both sides of (7.1) and using
the relation $d^2 = 0$ are called the Bianchi identities. Let

us derive them.

$$\begin{aligned}
d\underset{\sim}{\Omega} &= d(d\underset{\sim}{\omega}) + \frac{1}{2} d[\underset{\sim}{\omega},\underset{\sim}{\omega}] \\
&= 0 + \frac{1}{2}[d\underset{\sim}{\omega},\underset{\sim}{\omega}] - \frac{1}{2}[\underset{\sim}{\omega},d\underset{\sim}{\omega}] \\
&= \frac{1}{2}\left[\underset{\sim}{\Omega} - \frac{1}{2}[\underset{\sim}{\omega},\underset{\sim}{\omega}],\underset{\sim}{\omega}\right] - \frac{1}{2}\left[\underset{\sim}{\omega},\underset{\sim}{\Omega} - \frac{1}{2}[\underset{\sim}{\omega},\underset{\sim}{\omega}]\right] \\
&= \frac{1}{2}([\underset{\sim}{\Omega},\underset{\sim}{\omega}] - [\underset{\sim}{\omega},\underset{\sim}{\Omega}]) + \frac{1}{4}([[\underset{\sim}{\omega},\underset{\sim}{\omega}],\underset{\sim}{\omega}] - [\underset{\sim}{\omega},[\underset{\sim}{\omega},\underset{\sim}{\omega}]])
\end{aligned} \qquad (7.2)$$

Exercise. Show that

$$[[\underset{\sim}{\omega},\underset{\sim}{\omega}],\underset{\sim}{\omega}] = 0 \qquad (7.3)$$

(This is a consequence of the Jacobi identity for the Lie algebra $\underset{\sim}{G}$.)

Combining (7.2) and (7.3) we have:

$$2d\underset{\sim}{\Omega} = [\underset{\sim}{\Omega},\underset{\sim}{\omega}] - [\underset{\sim}{\omega},\underset{\sim}{\Omega}] \qquad (7.4)$$

This is the Bianchi-identity.

As a consequence notice that:

> The exterior derivatives of the components of $\underset{\sim}{\Omega}$ lie in the Grassman algebra ideal of forms generated by the components of $\underset{\sim}{\Omega}$.

This is a generalization of the "Frobenius complete integrability condition" for one-forms. We shall say that a set of two-forms with this property <u>satisfies the Frobenius condition</u>. Wahlquist and Estabrook have verified by explicit calculation that many of the partial differential equations of non-linear wave theory can be defined by exterior differential systems with this property. This is another clue that tells us that their "prolongation" process is closely linked with the theory of Cartan-Ehresmann connections.

8. QUADRATIC PROLONGATIONS OF NON-LINEAR WAVE EQUATIONS IN TWO INDEPENDENT VARIABLES

Let us return to the explicit problem that motivated this excursion into the theory of connections--the Estabrook-Wahlquist prolongation structure of certain non-linear wave equations in two independent variables (the Korteweg-de Vries, Sine-Gordon, non-linear Schrödinger,...). I can <u>sketch</u> how the geometric ideas of connection theory can be applied.

Let (x,t) denote space-time variables, the "independent variables" for the partial differential equations. Let u denote the dependent variables. (The equations may involve vector-valued functions of (x,t).)

We suppose that X is a fiber space, with base space the space of variables (x,t,u). We also assume that there is a differential ideal I of two-forms on X with the following properties:

a) The two-dimensional solution submanifolds of I which are transversal to (x,t) (i.e., on which (x,t) can be introduced as independent variables) are, when projected down to (x,t,u)-space <u>exactly</u> the graph of maps $(x,t) \to u(x,t)$ which are solutions of the given partial differential equations.

b) I is generated by a set of two-forms $\alpha_1, \alpha_2, \ldots$

c) These two forms $\alpha_1, \alpha_2, \ldots$ have the Frobenius proper i.e., I is the Grassman algebra ideal generated by $\alpha_1, \alpha_2, \ldots$

With these facts as given (and they can be readily verified using Estabrook and Wahlquist's calculations) we see (now that it is a very natural mathematical idea to attempt to construct a fiber space area X, and a connection for this fiber space, such that the α's are expressible in terms of the curvature forms of this connection. Perhaps the simplest nontrivial example of such a connection would be that on a fiber space with R as fiber, and with $SL(2,R)$ as structure group acting via "linear fractional transformations" on R. We shall see that this leads to the "quadratic, one-pseudo potential" prolongations of Estabrook, Wahlquist and Corones.

Let $\underset{\sim}{G}$ be the Lie algebra of $SL(2,R)$. It is a three-dimensional real Lie algebra. Let Y be R, defined by the variable R. $\underset{\sim}{G}$ has a basis W_1, W_2, W_3. It is readily verified that $\underset{\sim}{G}$ acts as a Lie algebra of vector fields on Y as follows:

$$W_1 = \frac{\partial}{\partial y} \; ; \quad W_2 = y \frac{\partial}{\partial y} \; ; \quad W_3 = y^2 \frac{\partial}{\partial y} \; .$$

The key reason for this was first explained by Lie; up to local equivalence there is but one way for a <u>finite dimensional</u> Lie algebra to act on a one-dimensional Lie algebra. See "Sophus Lie's 1880 Transformation Group Paper" in the "Lie Group" series.)

Consider $X \times Y$ as a fiber space over X, and a Cartan-Gresmann connection with structure group G. We see that the connection is determined by a single one-form in $X \times Y$ which <u>quadratic</u> in y, i.e., it can be written as:

$$\boxed{\omega = \omega_0 + y\omega_1 + y^2\omega_2} \tag{8.1}$$

Conversely, any one-form of this type defines an $SL(2,R)$-connection in the fiber space.

<u>Definition</u>. This connection is said to be <u>associated with the differential equation</u> if its connection forms generate the ideal which defines the differential equation, and which has the properties a), b), c) described above.

I would now <u>conjecture</u> that such connections are "intrinsically" attached to the differential equation, in the sense that one equation can be obtained from another by a "geometric" transformation if and only if the connections are equivalent. This will be treated later on. There are many examples in Cartan's Collected Works of such connections "attached" to differential equations, although none of his examples--to my knowledge are those of the type considered by Estabrook and Wahlquist.

Using formula (4.3), we can write the conditions relating the connection form ω (i.e., the "pseudoconservation law" (8.1)) and the ideal I as follows:

$$\alpha_0 = d\omega_0 - \omega_0 \wedge \omega_1$$

$$\alpha_1 = d\omega_1 - 2\omega_0 \wedge \omega_2$$

$$\alpha_2 = d\omega_2 + \omega_1 \wedge \omega_2$$

generate the ideal I.

9. THE GEOMETRIC FOUNDATION OF THE INVERSE SCATTERING TECHNIQUE

The "inverse scattering" technique associates with certain non-linear partial differential equations other <u>linear</u> equations

PROLONGATION STRUCTURE

Standard ideas of "scattering theory" are applied to the linear equations, and consequences are deduced for the non-linear ones with which we started. So far, the starting point of the process--the actual construction of the linear equations--has been a hit-or-miss affair; Wahlquist and Estabrook's work gives an idea about systematizing the search, and I will now show what it means in terms of the theory of connections.

Let us start off in the following way. X is a manifold, Y a one-dimensional space, whose variable is denoted by y. Suppose given an $SL(2,R)$-connection in $X \times Y$, defined by a one-form on $X \times Y$ of the form:

$$\omega = dy + \omega_0 + y\omega_1 + y^2 \omega_2 \tag{9.1}$$

Set:

$$\begin{aligned} \Omega_0 &= d\omega_0 - \omega_0 \wedge \omega_1 \\ \Omega_1 &= d\omega_1 + 2\omega_0 \wedge \omega_1 \\ \Omega_2 &= d\omega_2 + \omega_1 \wedge \omega_2 \end{aligned} \tag{9.2}$$

The forms Ω_1, Ω_2, Ω_3 are the <u>curvature forms</u> of the connection.

We can now construct a <u>linear</u> fiber space structure over X by the classical trick of introducing homogeneous coordinates (z^1, z^2)

$$\text{with } y = \frac{z^2}{z^1} \tag{9.3}$$

Suppose given a connection for this fiber space with structure group $SL(2,R)$ (associated with the "linear" action of $SL(2,R)$ on R^2). It is determined by a set (ω^1, ω^2) of one-forms of the following type:

$$\omega^1 = dz^1 + z^1 \omega^1_1 + z^2 \omega^1_2$$

$$\omega^2 = dz^2 + z^1 \omega^2_1 + z^2 \omega^2_2$$

(9.4)

The matrix

$$\begin{pmatrix} \omega^1_1, & \omega^1_2 \\ \omega^2_1, & \omega^2_2 \end{pmatrix}$$

of one-forms also satisfies the following condition:

$$\omega^1_1 + \omega^2_2 = 0 \ .$$

(9.

We can now determine the connection (9.1) in terms of the connection (9.4). First, apply d to both sides of (9.3):

$$dy = \frac{z^1 dz^2 - z^2 dz^1}{(z^1)^2}$$

$$= \text{, using (9.4),}$$

$$\frac{(\omega^2 - z^1 \omega^2_1 - z^2 \omega^2_2)}{y^1} - \frac{z}{z^1} (\omega^1 - z^1 \omega^1_1 - z^2 \omega^1_2)$$

$$= \frac{\omega^2 - y\omega^1}{z^1} - \omega_1^2 - y\omega_2^2 + y\omega_1^1 + (y)^2 \omega_2^1$$

Thus, if we <u>define</u> as:

$$\omega = \frac{\omega^2 - y\omega^1}{z^1} \tag{9.6}$$

we have the following relation:

$$\omega = dy + \omega_0 + \omega_1 y + \omega_2 y^2, \tag{9.7}$$

with

$$\boxed{\begin{aligned} \omega_0 &= \omega_1^2 \\ \omega_1 &= 2\omega_2^2 = -2\omega_1^1 \\ \omega_2 &= -\omega_2^1 \end{aligned}} \tag{9.8}$$

<u>Remark</u>. This is a special case of a general procedure--a "projective" connection is determined by a "linear" connection in a vector bundle whose fiber has one more dimension. See IM, Vol. 10.

Let us now rewrite the <u>linear</u> connection forms (9.4) in terms of ω_0, ω_1, ω_2, using relations (9.8):

$$\boxed{\begin{aligned} \omega^1 &= dz^1 - z^1 \frac{\omega_1}{2} - z^2 \omega_2 \\ \omega^2 &= dz^2 + z^1 \omega_0 + z^2 \frac{\omega_1}{2} \end{aligned}} \qquad (9.9)$$

Exercise. Show that the vanishing of the curvature forms Ω_0, Ω_1, Ω_2 defined by (9.2) is the necessary and sufficient condition for the complete integrability of the Pfaffian system ω^1, ω^2. In other words, show that

$$\boxed{d \begin{pmatrix} \omega^1 \\ \omega^2 \end{pmatrix} \text{ lies in the Grassman algebra ideal generated by } \omega^1, \omega^2, \Omega_0, \Omega_1, \Omega_2} \qquad (9.10)$$

Let I be the differential ideal of forms on X generated by $\Omega_0, \Omega_1, \Omega_2$. (Because of the Bianchi identities, we know in fact that I is identical to the <u>Grassman</u> algebra generated by $\Omega_0, \Omega_1, \Omega_2$.) Typically, the two-dimensional integral submanifold of I will be determined by a non-linear partial differential equation in two independent variables. (For example we have seen how to do this for the Korteweg-de Vries equation.)

Consider a two-dimensional integral submanifold of I. Restrict everything to this submanifold. <u>Because of</u> (9.10), the

PROLONGATION STRUCTURE   199

forms ω^1, ω^2 are <u>completely integrable</u>. The <u>linear</u> equations for z^1, z^2 obtained by setting (9.9) equal to zero should be the equations involved in the inverse scattering technique. (This observation is due to Wahlquist and Estabrook.)

Let us make this more explicit. For physical reasons, denote the independent variables by (x,t). Denote the variable of X by

$$(x,t,p,q,r)$$

(Thus, we suppose--as in the Korteweg-de Vries equation--that X is five dimensional.) We are given functions

$$p(x,t',q(x,t),r(x,t)) \tag{9.11}$$

which, when substituted, make $\Omega_0, \Omega_1, \Omega_2$ zero. Denote by $\omega_0', \omega_1', \omega_2'$ the one-forms in variables (x,t) obtained by substituting into $\omega_0, \omega_1, \omega_2$ the functions (9.11). Suppose that

$$\omega_0' = a_0(x,t)dx + b_0(x,t)dt$$
$$\omega_1' = a_1(x,t)dt + b_1(x,t)dt \tag{9.12}$$
$$\omega_2' = a_2(x,t)dx + b_2(x,t)dt$$

To solve $\omega^1 = \omega^2 = 0$, we are looking for functions $z^1(x,t)$, $z^2(x,t)$ such that

$$dz^1 = \frac{z^1}{2}(a_1 dx + b_1 dt) - z^2(a_2 dx + b_2 dt)$$
$$dz^2 = -z^1(a_0 dx + b_0 dt) - \frac{z^2}{2}(a_1 dx + b_1 dt) \quad . \tag{9.13}$$

These equations are (by construction) <u>completely integrable</u>.
We can solve them as follows:

> Choose arbitrarily $t \to x(t)$ as any function of t. Substitute into (9.13), obtaining a set of ordinary, <u>linear</u> differential equations:
>
> $$\frac{d}{dt} \underline{z}(t) = A(t)(\underline{z}(t)) \quad , \qquad (9.14)$$
>
> with
>
> $$z(t) = \begin{pmatrix} z^1(t,x(t)) \\ z^2(t,x(t)) \end{pmatrix} \in R^2 \quad .$$
>
> These linear equations determine <u>parallel transport</u> in this vector bundle with respect to the linear connection defined by (9.9). The "spectral theory" of the linear system (9.14) may be expected to bear some relation to the original <u>non-linear</u> differential equation.

This is clearly an area which needs much more development and elaboration. I will go into this at a later point.

10. THE BÄCKLUND TRANSFORMATION IN THE SENSE OF ESTABROOK AND WAHLQUIST

When one looks for the geometric source of the "soliton" phenomenon one quickly realizes that the 19-th century concept of a "Bäcklund transformation"--which has been forgotten in the 20-th century--is the key. Now, the "contact transformations" are really the broadest class of symmetry ideas which have achieved a recognized and conventional place in contemporary physics and engineering. The "Bäcklunds" are a generalization--one would expect that they are destined to explain new physical phenomena, e.g., the "solitons". It will be particularly interesting to see what role Bäcklund transformations play in quantum mechanics. (Remember that Einstein predicted that some mathematical property of non-linear field equations would be found to explain elementary particles!)

Wahlquist and Estabrook--in their basic paper [1], which we have already used extensively--have described how their "prolongation" structure may be used to describe the Bäcklund transformations. I will now briefly go over some of this material, in the context of connection theory.

As before, work with the simplest sort of $SL(2,R)$-connection. Let X be a manifold, with a differential ideal I of differential forms on X. Let Y be the real numbers, parameterized by the variable y, (which Estabrook and Wahlquist call a pseudopotential for I).

Suppose that a one-form

$$\omega = \omega_0 + y\omega_1 + y^2\omega_2 \tag{10.1}$$

is given on $X \times Y$, which defines an $SL(2,R)$-connection, whose curvature forms

$$\Omega_0 = d\omega_0 - \omega_0 \wedge \omega_1$$
$$\Omega_1 = d\omega_1 + 2\omega_0 \wedge \omega_2 \tag{10.2}$$
$$\Omega_2 = d\omega_2 + \omega_1 \wedge \omega_2$$

generate I.

A <u>symmetry</u> (in the traditional sense) of the differential equations associated with I may be defined as a diffeomorphism

$$\phi: X \to X$$

such that

$$\phi^*(I) = I \quad . \tag{10.3}$$

Condition (10.3) of course, guarantees that ϕ maps a solution submanifold of I into another solution submanifold. Given the relation between these solution submanifolds and the original set of differential equations, we see that ϕ determines a "symmetry" of the differential equation, in the sense that it maps solutions into solutions.

Here is the generalization:

Definition. A diffeomorphism

$$\beta: X \times Y \to X \times Y$$

is said to be a Bäcklund symmetry (in the sense of Wahlquist and Estabrook) if the following condition is satisfied:

$$\beta^*(I) \subset \text{Grassman ideal generated by } I \text{ and } \omega \qquad (10.4)$$

Notice particularly that the composition of two Bäcklunds is not necessarily a Bäcklund, since $\beta^*(\omega)$ is not required to be in the Grassman ideal generated by I and ω! Thus, these types of symmetries have a basically different algebraic structure (e.g., they should not be expected to form groups, at least in simple and natural ways) from the "symmetries" to which we have become accustomed in physics. This remark might be particularly important in physics, e.g., as a guide to the role of group theory in elementary particle physics. Perhaps groups as we know them are the appropriate algebraic structures only for "free" particles--interacting ones might involve something different!

We can now see how relation (10.4) is related to the way Bäcklund transformations may be expected to act on the solutions of differential equations. Let

$$\alpha: Z \to X$$

be a solution submanifold of I, i.e.,

$$\alpha^*(I) = 0 . \tag{10.5}$$

(In the usual example, $Z = R^2$, the space of variables (x,t)). Since I contains the curvature forms of ω, the map α can be lifted to give a map

$$\alpha': Z \to X \times Y ,$$

such that:

$$\alpha'^*(\omega) = 0 \tag{10.6}$$

$$\alpha = \pi\alpha' ,$$

where

$$\pi: X \times Y \to X$$

is the Cartesian projection. (In other words, α' is of the form

$$z \to (\alpha(z), y(z))$$

Apply β to α':

$$(\beta\alpha')^*(I) = (\alpha'^*)(\beta^*(I))$$

$$\leftarrow \subset (\alpha'^*)(I+\omega)$$

$$= 0, \text{ because of (10.5) } \underline{\text{and}} \text{ (10.6).}$$

In other words, β assigns to each <u>lifting</u> α', which satisfies (10.6), a solution submanifold α <u>another</u> solution submanifold

PROLONGATION STRUCTURE

205

$$\pi\beta\alpha': Z \to X \quad .$$

However, $\beta\alpha'$ does not necessarily satisfy (10.6), which is the fact which negates the group property of the symmetry. (Also, the lifting α' subject to (10.6) is not unique; the Bäcklund transformations really have a parameter intrinsically built in. Again, this property of such operations--which differs from the usual way symmetries have appeared in physics--might be useful in the study of elementary particles.

As examples of this phenomenon, I can refer to Wahlquist and Estabrook's work in [1], at the end of page 5 and the beginning of page 6. They show how these ideas generate the known Bäcklund transformations of the K-dV equation. Here, X has variables labelled

$$(x,t,u,p,z)$$

$$\beta: X \times Y \to X \times Y$$

is determined by condition (10.4) and the following formulas:

$$\boxed{\begin{aligned} \beta^*(x) &= x \, ; & \beta^*(t) &= t \, . \\ \beta^*(y) &= y \, ; & \beta^*(u) &= -u - y^2 + \lambda \end{aligned}} \quad (10.7)$$

(λ is a constant which is carried along in the definition of β.) They show (in formulas (5.8), (5.9)) how β may be applied to the "vacuum" solution, $u \equiv 0$, of the K-dV equation to give the "one-soliton" solution. It is this property of Bäcklund transformations--generating soliton solutions--that indicates their ultimate importance in the understanding of elementary particles!

Chapter VIII

BÄCKLUND TRANSFORMATIONS

1. THE BÄCKLUND TRANSFORMATION OF THE SINE-GORDON EQUATION

Here is the remarkable example which started the theory of Bäcklund transformations. Refer to Darboux' "Theorie des surfaces", Vol. 3, Section 808, formula 35.

$$\boxed{\begin{aligned} \partial_x z_1 + \partial_x z_2 &= a \sin(z_1 - z_2) \\ \partial_y z_1 - \partial_y z_2 &= b \sin(z_1 + z_2) \end{aligned}} \quad (1.1)$$

Here, (x,y) are independent variables. z_1, z_2 are functions of them. a and b are constants. (1.1) is to be considered as a system of partial differential equations for functions $z_1(x,y)$, $z_2(x,y)$.

There are compatibility conditions that one finds by differentiating the first line in (1.1) with respect to y, the second with respect to x.

$$\begin{aligned} \partial_{yx} z_1 + \partial_{yx} z_2 &= a \cos(z_1 - z_2)\partial_y(z_1 - z_2) \\ &= \underline{\text{using (1.1) again}}, \\ &\quad a \cos(z_1 - z_2)\, b \sin(z_1 + z_2) \quad (1.2) \end{aligned}$$

$$\partial_{xy}z_1 - \partial_{xy}z_2 = b \cos(z_1+z_2)\partial_x(z_1+z_2)$$

$$= b \cos(z_1+z_2) a \sin(z_1-z_2) \qquad (1.3)$$

Now, add (1.2) and (1.3):

$$2\partial_{xy}z_1 = ab(\cos(z_1-z_2)\sin(z_1+z_2) + \cos(z_1+z_2)\sin(z_1-z_2))$$

= , using the addition formula for trigonometric functions,

$$ab \sin(2z_1) \ .$$

This is a <u>differential equation for</u> z_1 alone. Similarly, we can subtract (1.3) from (1.2), obtaining the same differential equation for z_2.

We can now choose a and b appropriately, and change the independent variables by a linear combination, which we label for obvious physical reasons

$$(x,t) \ ,$$

i.e., <u>space</u> and <u>time</u>, so that this differential equation become the following one:

$$\boxed{\frac{\partial^2 z}{\partial t^2} - \frac{\partial^2 z}{\partial x^2} = \sin z} \qquad (1.$$

BÄCKLUND TRANSFORMATIONS

This equation is now called the Sine-Gordon equation, and has played a prominent role in recent work in engineering, physics, and applied mathematics. We see that the formulas (1.1) set up a "correspondence" (but not a mapping) between solutions of (1.4). This phenomena was, in the 19-th century, given the name Bäcklund transformation. They seem to be the geometric key to understanding the meaning of "solitons". In this chapter I will present a general setting for the concepts, and discuss other examples.

2. DIFFERENTIAL EQUATION HOMOMORPHISMS AND BÄCKLUND TRANSFORMATIONS

Let us now use the "mapping element" concept described in Part A. Let X and Z be manifolds; X is the space of independent variables, Z the space of dependent variables. $M^r(X,Z)$ denotes the space of r-th order mapping elements. $M(X,Z)$ denotes the space of maps

$$\underline{z}: X \to Z \quad .$$

Each $\underline{z} \in M(X,Z)$ defines a map

$$\partial^r \underline{z}: X \to M^r(X,Z) \quad .$$

A differential equation,

$$DE \subset M^r(X,Z)$$

is a submanifold of the mapping element space. $\underline{z} \in M(X,Z)$ is a <u>solution</u> of the differential equation if:

$$\partial^r \underline{z}(X) \subset DE .$$

The space of all solutions is denoted by

$$\underline{DE} .$$

Now, let Z, Z' be two dependent variables manifolds. Le

$$\phi: Z' \to Z$$

define a map. ϕ defines a map, denoted by

$$\underline{\phi}: M(X,Z') \to M(X,Z) ,$$

by composition:

$$\underline{\phi}\, z'(x) = \phi(z'(x))$$

for $z' \in M(X,Z')$.

<u>Definition</u>. Let DE, $(DE)'$ be two differential equations, one with Z as dependent variable manifold, the other with both with X as independent variable manifold. ϕ defines a <u>homomorphism</u> from $(DE)'$ to (DE) if the following conditio is satisfied:

$$\boxed{\underline{\phi}((\underline{DE})') \subset \underline{DE}} \qquad (2.$$

BÄCKLUND TRANSFORMATIONS

In words, if $x \to \underline{z}'(x)$ is a solution of one differential equation, its image

$$x \to \phi(\underline{z}'(x))$$

must also be a solution.

Remark. This is not at all the most general concept of "homomorphism" for differential equations, but is the one which is adequate for our immediate purposes.

Definition. Let DE, (DE)" be differential equations, with dependent variable manifolds Z and Z", and with common independent variable manifold X. A Bäcklund transformation between DE and (DE)" is defined by giving the following data:

  a) A dependent variable manifold Z',

  b) maps $\phi: Z' \to Z$, $\phi'': Z' \to Z''$

  c) a differential equation (DE)' with Z' as dependent variable manifold.

$\underline{\phi}$ and $\underline{\phi}''$ then should be homomorphisms from the differential equations (DE)' to (DE) and (DE)", respectively. (DE)' is called the interlocking differential equation.

Let us review the meaning of this definition in more concrete terms. Let

$$\underline{z}': X \to Z'$$

be a map which is a solution of the interlocking differential equation. Then,

$$\underline{z} = \phi(\underline{z}')$$

is a solution of DE.

$$\underline{z}'' = \phi''(\underline{z}')$$

is a solution of (DE)".

Thus, we have established some sort of "correspondence"

$$\underline{z} \to \underline{z}''$$

between solutions of DE and solutions of (DE)". (Of course, this is not a "mapping" between solutions, in the usual sense that "mapping" is now used in mathematics--to one solution $\underline{z}$ of DE may correspond many solutions of (DE)". It is, howeve something like the idea of "algebraic correspondence" in algebraic geometry. It is also reminiscent of E. Cartan's noti of "equivalence" between geometric structures. It would be interesting to find out whether Cartan's definition is broad enough to cover Bäcklund's. (I do not recall reading anywhere in Cartan's works that he had this in mind, but perhaps the thought is implicitly there.)

We can now see how the classical example (described in Section 1) of the Bäcklund transformation between two copies of the Sine-Gordon equation fits into this framework. Here,

$X = R^2$ = space of variables (x,y),

$Z = Z'' = R$ = space of variable z,

$Z' = R^2$ = space of variables (z_1, z_2),

$\phi: Z' \to Z$, $\phi'': Z' \to Z''$,

are the Cartesian projections:

$$\phi(z_1, z_2) = z_1 \;;\quad \phi''(z_1, z_2) = z_2 \;.$$

The interlocking differential equation (DE) is that given by formula (1.1). DE and (DE)" are both the Sine-Gordon equation. The calculations given in Section 1 show that ϕ and ϕ'' are indeed "homomorphisms".

We will now examine another example.

3. LINEARIZATION OF BURGER'S EQUATION BY MEANS OF A BÄCKLUND TRANSFORMATION

This example is given in Whitham [1], p. 610. Here

$Z = R = Z'$

$Z' = R^2$ = space of variables z_1, z_2 .

The interlocking equations are:

$$\boxed{\begin{aligned}\partial_x z_2 &= a z_1 z_2 \\ \partial_y z_2 &= b z_2 \partial_x(z_1) + c z_1^2 z_2\end{aligned}} \qquad (3.1)$$

(a,b,c are constants). Let us try to <u>separate</u> equations for z_1, z_2 implied by these equations. In this case, this can be done readily by dividing both sides of (3.1) by z_2:

$$\partial_x(\log z_2) = az_1$$
$$\partial_y(\log z_2) = b\partial_x(z_1) + cz_1^2$$

Equate the cross derivatives

$$\partial_y \partial_x (\log z_2) = \partial_x \partial_y (\log z_2) \quad ,$$

or

$$a\partial_y(z_1) = b\partial_x^2(z_1) + 2cz_1 \partial_x(z_1) \quad (3.2)$$

This is <u>Burger's equation</u>.

We can find the equation to be satisfied by z_2 by applying ∂_x to both sides of the first equation of (3.1):

$$\partial_x^2 z_2 = a\partial_x(z_1)z_2 + za_1\partial_x(z_2)$$

$$= \frac{a}{b}(\partial_y(z_2) - cz_1^2 z_2) + az_1(az_1 z_2) \quad (3.3)$$

<u>Exercise</u>. Show that, in order that z_2 satisfy a differential equation involving <u>itself alone</u> and not z_1, the second and third terms on the right hand side of (3.3) must cancel.

BÄCKLUND TRANSFORMATIONS

The condition found in this exercise forces:

$$\frac{ac}{b} = a^2,$$

or

$$\boxed{c = ab} \qquad (3.4)$$

The equation for z_2 becomes:

$$\boxed{\partial_x^2 z_2 = \frac{a}{b} \partial_y z_2} \qquad (3.5)$$

This is the <u>heat equation</u>. We conclude that the Bäcklund transformation determined by formula (3.1) relates the Burger equation, which is <u>non-linear</u>, to the <u>heat equation</u>, which is, of course, linear.

Chapter IX

THE LAPLACE-DARBOUX TRANSFORMATION, LINEAR BÄCKLUND TRANSFORMATIONS, AND THE INVERSE SCATTERING TECHNIQUE

1. INTRODUCTION

The "Laplace transform" is, in the classical differential geometry literature, not what one means by that today, but rather a transformation process affecting linear partial differential equations in two independent variables. Its theory is most extensively developed in Volume 2 of Darboux' "Theorie des surfaces". To avoid confusion with the modern usage, I will give it the name "Laplace-Darboux".

It is basically a theory of "invariants" of linear differential operators under certain types of Bäcklund transformations. It is thus of great interest as an example of a topic that has been extensively worked out in a special case, but that might admit significant generalizations and applications to new areas.

One striking feature is the way the work presented by Darboux anticipates recent work on the Inverse Scattering Technique, i.e., the reciprocity between solutions of linear and non-linear partial differential equations.

2. THE GAUGE GROUP AND ITS DIFFERENTIAL INVARIANTS

The classical theory--treated in Chapter 2 of Volume 2 of Darboux--deals with linear second order differential operators of the form:

$$\Delta = \partial_{xy} + a\partial_x + b\partial_y + c \tag{2.1}$$

where a, b, c are scalar-valued functions of two independent variables (x,y). As in the classical literature, we shall let Δ act on scalar-valued functions

$$(x,y) \to z(x,y) \quad,$$

i.e., elements of

$$M(R^2, R) \quad.$$

We denote such a map by

$$\underline{z} \quad.$$

There are two groups acting on $M(R^2, R)$ which preserve this type of differential operator Δ:

a) The map $\underline{z} \to \lambda \underline{z}$; where $\lambda(x,y)$ is an arbitrary scalar-valued function. These are called <u>pure gauge transformations</u>. They form a group denoted by G.

b) The diffeomorphism

$$\alpha: R^2 \to R^2$$

such that $\alpha^*(x)$ is a function $\phi(x)$ of x alone, $\alpha^*(y)$ is a function $\psi(y)$ of y alone. They form a group (isomorphic to DIFF(R) × DIFF(R), denoted by G_2.

Now, G_2 <u>normalizes</u> the action of G_1 on $M(R^2,R)$. Thus, the semi-direct product group

$$G = G_2 \cdot G_1$$

can be formed. It is a group (an "infinite Lie group", in the sense of Lie and Cartan) and the problem of finding its <u>differential invariants</u> can be posed. (See Volume 3 of the "Lie groups" series, i.e., the translation of Lie's 1884 Differential Invariants paper and Wilczynsky's book [1] for the most accessible and simple treatment of these ideas in the classical literature.) Darboux shows how two first-order differential invariants can be constructed from the differential operator (2.1). The first step is to calculate the transform of Δ under the action of G_1 and G_2.

First, consider the pure gauge transformation:

$$\underline{z} \to \lambda \underline{z} \equiv \underline{z}'\ .$$

$$\partial_x(\lambda \underline{z}) = \partial_x(\lambda)\underline{z} + \lambda \partial_x \underline{z}$$

$$\partial_{yx}(\lambda \underline{z}) = \partial_{yx}(\lambda)\underline{z} + \lambda\partial_{yx}(\underline{z}) + \partial_y(\lambda)\partial_x(\underline{z}) + \partial_x(\lambda)\partial_y\underline{z}$$

Hence,

$$\Delta(\lambda\underline{z}) = \lambda\partial_{xy}(\underline{z}) + \partial_x(\lambda)\partial_y\underline{z} + \partial_y(\lambda)\partial_x\underline{z} + z\partial_{xy}(\lambda)$$
$$+ a(\partial_x(\lambda)\underline{z} + \lambda\partial_x\underline{z}) + b(\partial_y(\lambda)\underline{z} + \lambda\partial_y\underline{z}) + c\lambda\underline{z}$$

Hence,

$$\lambda^{-1}\Delta(\lambda\underline{z}) = \partial_{xy}(\underline{z})$$
$$+ (\lambda^{-1}\partial_y(\lambda) + a)\partial_x\underline{z}$$
$$+ (\lambda^{-1}\partial_x(\lambda) + b)\partial_y\underline{z}$$
$$+ (\lambda^{-1}\partial_{xy}(\lambda) + a\partial_x(\lambda) + b\partial_y(\lambda) + c)z \quad,$$

or

$$\boxed{\begin{aligned}\Delta \to \lambda^{-1}\Delta\lambda \equiv{}& \partial_{xy} + (\lambda^{-1}\partial_y(\lambda) + a)\partial_x \\ & + (\lambda^{-1}\partial_x(\lambda) + b)\partial_y \\ & + (\lambda^{-1}\partial_{xy}(\lambda) + a\lambda^{-1}\partial_x(\lambda) + b\lambda^{-1}\partial_y(\lambda) + c)\end{aligned}}$$

(2.2)

Formula (2.2) determines the action of G_1 on linear differential operators of the form (2.1).

LAPLACE-DARBOUX

In order to construct differential invariants of G_1, set

$$\boxed{\begin{aligned} h(\Delta) &= \partial_x(a) + ab - c \\ k(\Delta) &= \partial_y(b) + ab - c \end{aligned}} \qquad (2.3)$$

Then, using (2.2),

$$\begin{aligned} h(\lambda \Delta \lambda^{-1}) &= \partial_x(\lambda^{-1}\partial_y\lambda + a) \\ &\quad + (a + \lambda^{-1}\partial_y\lambda)(b + \lambda^{-1}\partial_x(\lambda)) \\ &\quad - (\lambda^{-1}\partial_{xy}(\lambda) + a\lambda^{-1}\partial_x(\lambda) + b\lambda^{-1}\partial_y(\lambda) + c) \\ &= -\lambda^{-2}\partial_x(\lambda)\partial_y(\lambda) + \lambda^{-1}\partial_{xy}\lambda + \partial_x(a) \\ &\quad + ab + b\lambda^{-1}\partial_y\lambda + a\lambda^{-1}\partial_x(\lambda) + \lambda^{-2}\partial_y(\lambda)\partial_x(\lambda) \\ &\quad - \lambda^{-1}\partial_{xy}(\lambda) - a\lambda^{-1}\partial_x(\lambda) - b\lambda^{-1}\partial_y(\lambda) - c \\ &= h(\Delta) \quad . \end{aligned}$$

This shows that:

$$\boxed{h \text{ is invariant under the action of } G_1.}$$

Similarly, one can show that

> k, as a map of
>
> (differential operators) → (functions)
>
> is invariant under the action of G_1.

One can now readily calculate how the functions h,k change under G_2.

<u>Exercise</u>. Let $\alpha: R^2 \to R^2$ be a diffeomorphism of the space of independent variables which is of the form

$$\alpha^{-1}(x,y) = (\phi(x), \psi(y))$$

Show that

$$\alpha^{-1} h(\alpha(\Delta)) = \phi\psi h(\Delta)$$

$$\alpha^{-1} k(\alpha(\Delta)) = \phi\psi k(\Delta)$$

(2.4)

In words, the functions h,k are what are called in the classical literature, <u>relative invariants</u> of the group G_2, i.e., transform by a factor.

3. THE LAPLACE-DARBOUX TRANSFORM AS A LINEAR BÄCKLUND TRANSFORMATION

Introduce the usual coordinates for $M^1(R^2, R)$:

$$p = \partial_x z ; \quad q = \partial_y z .$$

Recall (from the previous chapter) that a Bäcklund transformation involves interlocking equations of the form:

$$f(x,y,z,z_1,p,q,p_1,q_1) = 0$$

$$g(x,y,z,z_1,p,q,p_1,q_1) = 0 \tag{3.1}$$

Two differential equations $\Delta = 0$, $\Delta_1 = 0$ are Bäcklund transforms of each other if the solutions

$$(x,y) \to (z(x,y), z_1(x,y)) \equiv (\underline{z}, \underline{z}_1)$$

of (3.1) have the property that:

$$\Delta(\underline{z}) = 0$$

$$\Delta_1(\underline{z}_1) = 0 .$$

(In practice, one often replaces this condition by a slightly stronger one, requiring that Δ, Δ_1 be the integrability conditions of the Mayer systems resulting from (3.1).)

Suppose that we start off with Δ as the linear, second order differential operator of the form (2.1). Let us also suppose that the equations (3.1) are linear in z, z_1, p, q, i.e that they are of the following form:

$$\boxed{\begin{aligned} z_1 - \partial_y z - az &= 0 \\ \partial_x z_1 + b z_1 - \alpha z &= 0 \end{aligned}} \quad (3.2)$$

a,b are the coefficients of the differential operator Δ, as given by formula (2.1). α is a function <u>which must be determined</u> by the condition that (3.2) determines a Bäcklund transformation of $\Delta = 0$ to another linear, constant coefficient differential equation.

In order to express the Bäcklund property for (3.2) (i.e to show that (3.2) defines an interlocking system for two equations, one of which is (2.1)) let us apply ∂_x to the first term of (3.2):

$$\partial_x z_1 = \partial_{xy} z + \partial_x(a) z + a \partial_x(z) \quad (3.3)$$

Equate the right hand side of (3.3) to $\partial_x z_1$ as <u>determined b the second term</u> in (3.2):

$$\partial_{xy} z + \partial_x(a) z + a \partial_x(z) = -b z_1 + \alpha z$$

$$= \text{, using the first terms i (3.2) again,}$$

$$-b(\partial_y z + az) + \alpha z$$

LAPLACE-DARBOUX

or,

$$\partial_{xy}(z) + a\partial_x(z) + b\partial_y(z) + (\partial_x(a)+ba-\alpha)z = 0 \qquad (3.4)$$

We see that, <u>in order that equation</u> (3.4) <u>coincide with equation</u> (2.1), we must have

$$\alpha = \alpha_x(a) + ba - c$$
$$= \quad , \text{ using } (2.3),$$
$$h(\Delta)$$

Let us sum up as follows:

Theorem 3.1. Let

$$\Delta(z) = \partial_{xy}z + a\partial_x z + b\partial_y z + cz = 0$$

be a linear, second order P.D.E. in two independent variables. Let $h(\Delta)$, $k(\Delta)$ be the invariants of this equation under the gauge group, as determined by formula (3.2). Then, the formula:

$$\begin{aligned} z_1 &= \partial_y z + az \\ \partial_x z_1 &= h(\Delta)z - bz_1 \end{aligned} \qquad (3.5)$$

$$\boxed{\begin{aligned} z_{-1} &= \partial_x z + bz \\ \partial_y z_{-1} &= k(\Delta)z + az_{-1} \end{aligned}} \quad (3.6)$$

define Bäcklund transformations relating Δ to linear differential operators Δ_1, Δ_{-1}, which are called the <u>Laplace-Darboux transforms of</u> Δ.

(That (3.6) also defines a Bäcklund transformation is proved in a similar way.)

<u>Theorem 3.2.</u> The linear differential operator Δ is transfor via the Bäcklund transformation defined by the interlocking equations (3.5) and (3.6), into the differential operators Δ, Δ_1 given by the following formulas:

$$\boxed{\begin{aligned} \Delta_1 = \partial_{xy} &+ (a - \partial_y / \log(\Delta))\partial_x + b\partial_y \\ &+ (c - \partial_x(a) + \partial_y b - b\partial_y(\log h(\Delta))) \end{aligned}} \quad (3.7)$$

$$\boxed{\begin{aligned} \Delta_{-1} = \partial_{xy} &+ a\partial_x + (b - \partial_x \log k(\Delta))\partial_y \\ &+ (c - \partial_y(b) + \partial_x(a) - a\partial_x(\log k(\Delta))) \end{aligned}} \quad (3.8)$$

LAPLACE-DARBOUX

The invariants of these transformed operators are given by the following formulas:

$$h(\Delta_1) = 2h(\Delta) - k(\Delta) - \partial_{xy}(\log h(\Delta)) \quad (3.9)$$

$$k(\Delta_1) = h(\Delta) \quad (3.10)$$

$$h(\Delta_{-1}) = k(\Delta) \quad (3.11)$$

$$k(\Delta_{-1}) = 2k(\Delta) - h(\Delta) - \partial_{xy}(\log k(\Delta)) \quad (3.12)$$

The proof of these formulas is given in Darboux, Vol. II, pp. 27-28. They are left as <u>Exercises</u>.

Thus we have an interesting situation: a group G acting on the set of differential operators; "semi-invariants"

$$\Delta \to h(\Delta), \; k(\Delta)$$

which are reproduced by a factor under the group; finally, the two Bäcklund transformations

$$\Delta \to \Delta_1, \; \Delta_{-1}$$

which act on the semi-invariant functions h, k as <u>linear differential operators</u>. This is the only place in the classical literature where this sort of structure is encountered, but if it could be generalized to apply to other, more complicated, situations, it might give a significant insight into the theory of partial differential equations.

Of course, Darboux himself has extensively worked out the implications of this structure for this specific class of partial differential equations, with emphasis on the problems encountered in the applications to classical surface theory. There is an enormous amount of such material in "Theorie des Surfaces"--particularly in Volume II, but one will find more in other volumes as well. I would suggest that someone gather it all together in a modern form, and see what can be generalized!

As an illustration of why this might be worthwhile, I will describe some material (pages 31-32 of Volume II) which seems to anticipate the Inverse Scattering Technique.

4. SOME PHYSICAL INTERPRETATIONS OF DARBOUX' WORK. THE KLEIN AND SINE-GORDON EQUATIONS

The idea of the Inverse Scattering Technique (see Lax [1], Witham [1]) seems to be to consider classes of <u>linear</u> differential operators whose coefficients satisfy certain <u>non-linear</u> partial differential equations. The simplest example (discovered by M. Kruskal and coworkers) seems to be the case where the linear equation is the Schrödinger equation and the non-linear one is the Korteweg-de Vries equation.

This correspondence <u>can in favorable cases</u> be exploited
in the following way: Classical <u>scattering theory</u> (particularly,
the work of Gelfand and Levitan [1] on "inverse scattering")
enables one to set up a correspondence between the coefficients
of the linear differential operators and certain "scattering
data". Since the coefficients are solutions of a fixed non-
linear partial differential equation, there is a correspondence--
which can be written in terms of integral equations using Gelfand-
Levitan--between solutions of the non-linear partial differential
equation and the scattering data. In the favorable cases this
can be exploited to find explicit solutions of the non-linear
partial differential equation.

So far, success of this method is hit-or-miss (and restricted
to partial differential equations in two independent variables).
I believe that the key to systematization and generalization
(and perhaps finding the true mathematical and physical reason
for the applicability of the method) is understanding the <u>differ-
ential-geometric</u> meaning of the correspondence between linear
and non-linear partial differential equations. Thus, it was
very interesting for me to find that Darboux had already encountered an example of this correspondence, which I will now describe.
(Of course, for quite different reasons the classical geometers
had already encountered soliton-type solutions of certain non-
linear partial differential equations, e.g., the Sine-Gordon
equation. This arises in connection with what they called
Transformation and Deformation of Surfaces.)

Start off with a single linear differential operator

$$\Delta = \partial_{xy} + a\partial_x + b\partial_y + c \qquad (4.1)$$

Applying the Laplace-Darboux transforms to it, we obtain two new differential operators

$$\Delta_1, \Delta_{-1},$$

given by formulas described in Section 3. The transforms can be iterated, obtaining operators

$$(\Delta_1)_1, \quad (\Delta_1)_{-1}, \quad (\Delta_{-1})_1, \quad (\Delta_{-1})_{-1},$$

and so on.

We shall now show that $(\Delta_1)_{-1}$ and $(\Delta_{-1})_1$ are not "new". To see this use formulas (3.5) and (3.6), which may be interpreted as determining the transformations as acting on the dependent variables:

$$(z_{-1})_1 = \partial_y z_{-1} + a z_{-1}$$

$$= \partial_y(\partial_x z + bz) + a(\partial_x z + bz)$$

$$= \partial_{xy} z + \partial_y(b) z + b\partial_y x + a\partial_x z + abz$$

$$= \text{, using the fact that } z \text{ satisfies } \Delta z = 0,$$

$$(\partial_y(b) + ab - c)$$

$$= k(\Delta) z$$

LAPLACE-DARBOUX

Similarly, we have:

$$(z_1)_{-1} = h(\Delta)z$$

These formulas prove the following:

Theorem 4.1.

$$(\Delta_{-1})_1 = k(\Delta)^{-1}\Delta k(\Delta)^{-1} \qquad (4.4)$$

$$(\Delta_1)_{-1} = h(\Delta)^{-1}\Delta h(\Delta)^{-1} \qquad (4.5)$$

Here is the significance of these formulas, as explained by Darboux. Consider the "pure" gauge transformations G_1 acting on the space of differential operators of the form (4.1). Let $\underset{\sim}{D}$ be the <u>orbit space</u> of this action. Consider the iteration of the operations

$$\begin{aligned}\Delta &\to \Delta_1 \\ \Delta &\to \Delta_{-1}\end{aligned} \qquad (4.6)$$

Exercise. Show that the maps (4.6) on differential operators map orbits of G_1 into orbits, i.e., pass to the quotient to act on the orbit space $\underset{\sim}{D}$.

Equations (4.4) and (4.5) now tell us that the operations (4.6) are, <u>when acting on</u> $\underset{\sim}{D}$, inverses of each other. Thus, the Bäcklund transformations of the Laplace-Darboux type form, when considered as acting on the orbit space $\underset{\sim}{D}$, an infinite cyclic group. Darboux denotes the iterates of an operator Δ under this action in the following obvious way:

$$\boxed{\begin{array}{l} \Delta_2 = (\Delta_1)_1, \ \Delta_3 = (\Delta_2)_1, \ldots \\ \\ \Delta_{-2} = (\Delta_{-1})_1, \ldots \end{array}} \qquad (4.7)$$

The sequence $\Delta_1, \Delta_2, \ldots, \Delta_{-1}, \Delta_{-2}, \ldots$ is called the Laplace-Darboux sequence.

Another interesting feature of the calculation--from the point of view of generalization and further application--is that the "invariants" $\Delta \to h(\Delta), k(\Delta)$ associated with the operator Δ are generated by formulas (4.4), (4.5) by means of the Bäcklund transformation.

Having defined the sequences of differential operators (4.7), and having seen that it has a natural geometric and invariant theoretic significance, one might expect that one can characterize various interesting classes of differential operators by means of their behavior relative to the Laplace-Darboux sequence (4.7). There is extensive work on this topic in Volume 2 of his treatise. The most immediate question of this type is:

LAPLACE-DARBOUX

> What are the properties of Δ which assure that the sequence (4.7) is periodic of period one or two.

Suppose first that the period is one, i.e.,

$$\Delta_1 = \Delta \;. \tag{4.8}$$

Hence,

$$\begin{aligned} k(\Delta) &= k(\Delta_1) \\ h(\Delta) &= h(\Delta_1) \end{aligned} \tag{4.9}$$

Remark. Keep in mind that the "equality" (4.8) means that the orbit in $\underset{\sim}{D}$ to which they belong is equal. Hence, the "equality" in (4.9) follows from the fact that $\Delta \to h(\Delta), k(\Delta)$ are the "invariants" of the action of G_1, the group of gauge transformations.

Exercise. Show that conversely, (4.9) implies (4.8), i.e., that if the invariants of Δ, Δ_1 are equal then they differ by a gauge transformation.

It is now a consequence of the Bäcklund transformation formulas (3.9)-(3.10) that

$$h(\Delta) = k(\Delta) \;. \tag{4.10}$$

In turn, Darboux shows (on p. 3), Section 337 of Vol. 2) that condition (4.10) implies that Δ can be chosen (after the action of the <u>full</u> gauge group $G = G_1 \cdot G_2$) to be of the form:

$$\Delta = \partial_{xy} - 1 \,. \tag{4.11}$$

This is essentially just the <u>Klein-Gordon</u> equation (in one space variable). Notice that it has been characterized by a simple and natural mathematical condition. Perhaps this is the reason for its relevance in Elementary Particle Physics! (In fact, in this case, the Gauge Transformations G_1 have a natural physical meaning--they lead to Electric Charge.) The group G_2 is also natural physically--since x,y are <u>really</u> "characteristic", i.e., of the coordinate form

$$\vec{x} \pm t \,,$$

where $\vec{x}, t$ are space-time variables, G_2 is the group on R^2 which preserves the "light cone" structure.

Now we can consider the case where the Laplace-Darboux sequence (4.7) is periodic of period <u>two</u>. This will lead us to the <u>Sine-Gordon equation</u>.

Use formulas (3.9)-(3.12). Let us assume that:

$$\boxed{\begin{aligned} h(\Delta_2) &= h(\Delta) \\ k(\Delta_2) &= k(\Delta) \end{aligned}} \tag{4.12}$$

LAPLACE-DARBOUX

Now, using (3.10),

$$k(\Delta_2) = h(\Delta_1)$$

$$= \text{, using (3.9)},$$

$$2h(\Delta) - k(\Delta) - \partial_{xy}(\log h(\ ))$$

$$= \text{, using (4.12)},$$

$$k(\Delta) \ .$$

Hence,

$$2h(\Delta) - 2k(\Delta) - \partial_{xy}(\log (h(\Delta))) = 0 \qquad (4.13)$$

Similarly, exploiting the fact that

$$h(\Delta_{-2}) = h(\Delta) \ ,$$

we have the symmetric formula:

$$2h(\Delta) - 2k(\Delta) - \partial_{xy} \log (h(\Delta)) = 0 \qquad (4.14)$$

(See Darboux, p. 32, Section 337 of Vol. 2.) Adding (4.13) and (4.14), gives:

$$\partial_{xy} \log (h(\Delta)) + \log (k(\Delta)) = 0 \ ,$$

or

$$\partial_{xy} \log (h(\Delta)k(\Delta)) = 0$$

or

$$h(\Delta)k(\Delta) = f(x)g(y) \tag{4.15}$$

Now we can change variables

$$(x,y) \to (x'(x), y'(y)) \quad ,$$

i.e., apply the group G_2. Since $h(\Delta)$ and $k(\Delta)$ change as semi-invariants under the group, i.e., multiply by a factor, we can suppose that coordinates are chosen so that:

$$h(\Delta)k(\Delta) = 1 \quad . \tag{4.16}$$

With this normalization, $h(\Delta)$ is the solution of the following equation:

$$\partial_{xy} \log (h(\Delta)) = 2h(\Delta) - 2h(\Delta)^{-1} \quad . \tag{4.17}$$

Set:

$$\psi = e^{ih(\Delta)} \tag{4.18}$$

Note that ψ satisfies:

$$\partial_{xy}\psi = c \sin \psi \quad , \tag{4.19}$$

for some constant c. This is the <u>Sine-Gordon equation</u>.

We see now how the Sine-Gordon equation determines operators Δ which transform with period two under the Sine-Gordon equation. A solution ψ of (4.19) determines $h(\Delta)$ via (4.18) then determine the coefficients a,b of Δ via the equations (2.3), i.e.,

$$\boxed{\begin{aligned} \partial_x(a) &= h(\Delta) + ab - c \\ \partial_y(b) &= h(\Delta)^{-1} + ab - c \end{aligned}} \qquad (4.20)$$

Notice how differently the Sine and Klein-Gordon equations appear from this Bäcklund transformation point of view. The non-linearity of the Sine-Gordon leads to much more intricate relations!

Another important observation is the way (4.18) exhibits ψ, the solution of the Sine-Gordon, as $e^{i(\text{something})}$. I believe that the proper generalization of these ideas to differential equations with more independent variables will involve mapping

$$\psi: R^4 \to (\text{a compact Lie group}) \ .$$

(This also fits in with some of Finkelstein's ideas [1] concerning the way the topological ideas will appear in elementary particle physics.)

Finally, note the relation to my version of WKB presented in VB, Vol. 2. Given a Schrödinger wave function $\psi(x,t)$, multiply it by a "gauge" transformation

$$\psi \to e^{iS/\hbar}\psi$$

The study of the transform of the Schrödinger operator under this guage transformation is closely linked to the physics of the situation.

Chapter X

HIGHER DERIVATIVE CONSERVATION LAWS FOR SYMPLECTIC MANIFOLDS

1. INTRODUCTION

As I have already emphasized, one encounters many diverse concepts in modern nonlinear wave theory. Obviously, one of the key ideas is that of <u>conservation law</u>. One of the first striking discoveries about "solitons" by Gardner, Green, Kruskal and Muira involved the discovery of an infinite number of such conservation laws for the Korteweg-de Vries equation involving <u>polynomials</u> in the unknown function, its higher derivatives, and the independent variables. Estabrook and Wahlquist have partially investigated what seems to be some sort of "algebraic structure" on such conservation laws. Gardner, Fadeev and Zhakharov have shown that these conservation laws define a canonical transformation on the infinite dimensional symplectic structure defined by the Korteweg-de Vries equation.

This circle of ideas clearly needs a general setting and systematization in terms of differential geometry. I hope to get into this in Part B. Here, I will briefly cover a restricted situation, investigating what happens for systems with a <u>finite number of degrees of freedom</u>, i.e., systems governed by ordinary differential equations.

Here is the general setting for such systems. Let x denote a vector in R^n, t a time variable. Consider a system of ordinary differential equations

$$\frac{dx}{dt} = f(x) \ . \tag{1.}$$

A function

$$(t,x) \to h(t,x)$$

is a <u>conservation law</u> for (1.1) if it is constant along each solution of (1.1). In terms of differential equations, this means that:

$$\partial_t h + \partial_x h(f(x)) = 0 \ . \tag{1.}$$

If the space of x's has a symplectic manifold structure and if the group generated by (1.1) is a one-parameter group of symplectic automorphisms, then the space of all conservation laws has a Lie algebra structure, since the Poisson bracket of two solutions of (1.2) is again a solution. This Lie algebra is obviously closely linked to quantum mechanics. (This is what I had in mind in titling my book "Lie Algebras and Quantum Mechanics", where the <u>field-theoretic</u> version of this material was developed, with continuation in VB, GPS, and IM, Vols. 5 and 6.)

There is obviously the possibility of considering conservation laws which depend on higher derivatives of x:

CONSERVATION LAWS

$$(x,t) \to (h(t,x,\dot{x},\ldots)) \; .$$

In turn, this can be considered as a conservation law of the prolonged system of (1.1), i.e., by adding to it the system obtained by differentiating (1.1):

$$\frac{d\dot{x}}{dt} = f_x(x)\dot{x}$$

$$\ldots \tag{1.3}$$

In this chapter I shall show that the symplectic structure can be extended to these higher derivative objects, providing a Poisson bracket structure for the higher derivative conservation laws. This will be done using the theory developed in Section 10, Chapter 12, Vol. 9 of IM.

2. THE SYMPLECTIC STRUCTURE ON THE TANGENT BUNDLE TO A SYMPLECTIC MANIFOLD

Let M be a manifold of $2n$ dimensions. A <u>symplectic structure</u> for M is defined by a two-differential form, such that:

$$d\omega = 0$$

$$\omega^n \equiv \omega \wedge \omega \wedge \cdots \wedge \omega \neq 0$$

Such an ω defines, for each point $p \in M$, an isomorphism between M_p and M_p^d (the <u>tangent</u> and <u>cotangent</u> spaces to M

at p), hence defines globally an isomorphism between the vector bundles $T(M)$ and $T^d(M)$ (the <u>tangent</u> and <u>cotangent</u> bundles). It is well known that $T^d(M)$ has a symplectic structure. Pulling this back via the isomorphism defined by ω defines a symplectic structure on $T(M)$. It will be called the <u>prolonged symplectic structure</u>. It was described in IM, Vol. 9, Chapter 12, Section 10, and plays an important role there in the differential geometry of networks (electrical and "generalized"). Here is how it is described in local coordinate

Choose indices and the summation convention as follows:

$$1 \leq i,j \leq n \ .$$

Let (p_i, q^j) be a <u>canonical</u> (local) <u>coordinate system</u> for M, i.e.,

$$\omega = dp_i \wedge dq^i \tag{2.1}$$

If $f \in F(M)$ is a real-valued C function on M, let

$$\dot{f}: T(M) \to R$$

be the function defined as follows:

$$\dot{f}(v) = df(v) \tag{2.2}$$

$$\text{for } v \in T(M) \ .$$

Then it is readily seen that

$$(p_i, q^j, \dot{p}_i, \dot{q}^j)$$

form a coordinate system for $T(M)$.

CONSERVATION LAWS

Theorem 2.1. The prolonged symplectic structure $\dot{\omega}$ on $T(M)$ is defined in these coordinates by these formulas:

$$\dot{\omega} = d\dot{p}_i \wedge dq^i + dp_i \wedge d\dot{q}^i \tag{2.3}$$

For the proof of (2.3), see page 405 of IM, Vol. 9.

Let us use (2.3) to construct the Poisson bracket operation on $T(M)$ defined by the symplectic form $\dot{\omega}$. Let $f(p,q,\dot{p},\dot{q})$ be a function on these coordinates. Let A_f be the vector field on $T(M)$ characterized by the following relations:

$$df = A_f \lrcorner \dot{\omega} . \tag{2.4}$$

Let us compute, using (2.3), an explicit formula for A_f.

$$A_f \lrcorner \dot{\omega} = A_f(\dot{p}_i)dq^i - A_f(q^i)d\dot{p}_i + A_f(p_i)d\dot{q}^i - A_f(\dot{q}^i)dp_i .$$

Compare this with the left hand side of (2.4), obtaining the following relations:

$$A_f(\dot{p}_i) = \frac{\partial f}{\partial q^i} \quad ; \quad A_f(q^i) = -\frac{\partial f}{\partial \dot{p}_i}$$

$$A_f(p_i) = \frac{\partial f}{\partial \dot{q}^i} \quad ; \quad A_f(\dot{q}^i) = -\frac{\partial f}{\partial p_i} \tag{2.5}$$

or

$$A_f = \frac{\partial f}{\partial q^i}\frac{\partial}{\partial \dot{p}_i} - \frac{\partial f}{\partial \dot{p}i}\frac{\partial}{\partial q^i} + \frac{\partial f}{\partial \dot{q}^i}\frac{\partial}{\partial p_i} - \frac{\partial f}{\partial p_j}\frac{\partial}{\partial \dot{q}^i} \qquad (2.6)$$

Hence, for $f_1, f_2 \in F(T(M))$,

$$\{f_1, f_2\} \equiv -A_{f_1}(f_2)$$

$$= \frac{\partial f_1}{\partial \dot{p}_i}\frac{\partial f_2}{\partial q^i} - \frac{\partial f_1}{\partial q^i}\frac{\partial f_2}{\partial \dot{p}_i} + \frac{\partial f_1}{\partial p_i}\frac{\partial f_2}{\partial \dot{q}^i} - \frac{\partial f_1}{\partial \dot{q}^i}\frac{\partial f_2}{\partial p_i} \qquad (2.7)$$

3. THE PROLONGATION FORMULA FOR VECTOR FIELDS

Keep the notation of Section 2. There is a map

$$V(M) \to V(T(M)) \quad,$$

called <u>prolongation</u>, which is geometrically (and "intrinsically") defined, denoted by

$$A \to \dot{A} \quad.$$

It can be described group theoretically as follows:

The one-parameter group of diffeomorphisms of $T(M)$ generated by $\dot{A}$ is the natural linear extension to tangent vectors of the one-parameter group of diffeomorphisms generated by A.

CONSERVATION LAWS

Analytically, it is characterized by the following relation:

$$\dot{A}(f) = \widehat{A(\dot{f})} \tag{3.1}$$

for $f \in F(M)$

$$\dot{A}(\pi^*(f)) = \pi^*(A(f)) \tag{3.2}$$

for $f \in F(M)$,

where $\pi: T(M) \to M$ is the projection map. Here is a main formula relating prolongation of vector fields and the prolongation of symplectic structure I defined in the previous section.

Theorem 3.1.

$$A_{\dot{f}} = \widehat{\dot{A}_f} \tag{3.3}$$

for $f \in F(M)$,

where $A_f, A_{\dot{f}}$ are the vector fields on M and T(M) defined as follows.

$$df = A_f \lrcorner \, \omega$$

$$d\dot{f} = \dot{A}_f \lrcorner \, \dot{\omega} \ .$$

Proof. Let $f(p,q)$ be the expression for f in the local canonical coordinates (p,q) for M. Then,

$$\dot{f} = \frac{\partial f}{\partial p_i} \dot{p}_i + \frac{\partial f}{\partial q^i} \dot{q}^i \tag{3.4}$$

Use (2.6):

$$A_{\dot{f}} = \frac{\partial \dot{f}}{\partial q^i}\frac{\partial}{\partial \dot{p}_i} - \frac{\partial f}{\partial p_i}\frac{\partial}{\partial q^i} + \frac{\partial f}{\partial q^i}\frac{\partial}{\partial p_i} - \frac{\partial \dot{f}}{\partial p_i}\frac{\partial}{\partial \dot{q}^i} \qquad (3.5)$$

Now,

$$A_f = \frac{\partial f}{\partial q^i}\frac{\partial}{\partial p^i} - \frac{\partial f}{\partial p_i}\frac{\partial}{\partial q^i}$$

Hence,

$$A_f(p_i) = \frac{\partial f}{\partial q^i} = A_{\dot{f}}(p_i)$$

$$A_f(q^i) = -\frac{\partial f}{\partial p_i} = A_{\dot{f}}(q^i)$$

$$A_f(\dot{p}_i) = \widetilde{\frac{\partial f}{\partial q_i}} = \frac{\partial \dot{f}}{\partial q_i} = A_{\dot{f}}(\dot{p}_i)$$

This proves (3.3).

<u>Theorem 3.2</u>. Let $h \in F(M)$ be a function on M, and let $t \to p(t)$ be a curve in M which is an orbit of the one-parameter group of symplectic automorphisms generated by h. Let

$$t \to \dot{p}(t)$$

be its tangent vector field. It is then an orbit of the one-parameter group of symplectic automorphisms generated by $\dot{h}$.

CONSERVATION LAWS

Proof. $t \to p(t)$ is characterized by the following equations

$$A_h \frac{d}{dt} f(p(t)) = A_h(f)(p(t))$$

for all $f \in F(M)$

Now

$$\dot{f}(\dot{p}(t)) = df(\dot{p}(t))$$

$$= \frac{d}{dt} f(p(t))$$

$$= A_h(f)(p(t)) \quad .$$

Hence

$$\frac{d}{dt} \dot{f}(\dot{p}(t)) = \widehat{A_h(f)}(\dot{p}(t))$$

$$= \text{, using (3.3) and (3.1)}$$

$$A_h^{\cdot}(\dot{f})(\dot{p}(t))$$

Also,

$$\frac{d}{dt} \pi^*(f)(\dot{p}(t)) = \frac{d}{dt} f(p(t))$$

$$= A_h(f)(p(t))$$

$$= A_h(f)(\pi(\dot{p}(t))$$

$$= \pi^*(A_h(f))(\dot{p}(t))$$

$$= \text{, using (3.2),}$$

$$\dot{A}_h(\pi^*(f)(\dot{p}(t))$$

$$= \text{, using (3.3),}$$

$$A_h^{\cdot}(\pi^*(f))(\dot{p}(t)) \quad .$$

These two calculations show that

$$t \to \dot{p}(t)$$

is an orbit curve of the vector field $A_h^{\cdot}$, completing the proof

<u>Corollary to Theorem 3.2</u>. If $t \to (p(t), q(t))$ is a solution to Hamilton equations

$$\frac{dq}{dt} = \partial_p h \quad , \qquad \frac{\partial p}{dt} = -\partial_q h \quad , \tag{3.6}$$

then the derivatives

$$t \to \left(p(t), \dot{p}(t) \equiv \frac{dp}{dt}, \; q(t), \dot{q}(t) \equiv \frac{dq}{dt} \right)$$

are solutions of the Hamilton equations, with Hamiltonian

$$\dot{h} \quad .$$

<u>Theorem 3.3</u>. If $(t,p) \to f(t,p)$ is a real valued function such that

$$\frac{\partial f}{\partial t} + A_h(f) = 0 \quad , \tag{3.7}$$

i.e., f is a <u>conservation law</u> for the Hamilton equations (3.6), then

$$\dot{f}$$

is a conservation law for the Hamilton equations with Hamiltonian $\dot{h}$.

Proof.

$$\frac{\partial \dot{f}}{\partial t} + A_{\dot{h}}(\dot{f}) = \frac{\partial \dot{f}}{\partial t} + \widehat{A_h}(\dot{f})$$

$$= \frac{\partial \dot{f}}{\partial t} + \widehat{A_h(f)}$$

$$= 0 ,$$

from (3.7).

Here is the main result from the point of view of "conservation laws".

<u>Theorem 3.4</u>. Let $h(p,q)$ be functions on M. A function $f(p,q,\dot{q},\dot{p},t)$ on $R \times T(M)$ is constant along all solutions of (3.6), i.e., is a <u>second order conservation law</u> for (3.6), if and only if

$$\frac{\partial f}{\partial t} + \{\dot{h},f\} = 0 . \qquad (3.8)$$

Proof.

$$\{\dot{h},f\} = -A_{\dot{h}}(f)$$

$$= \text{, using (3.3) again,}$$

$$-\widehat{\dot{A_h}}(f) .$$

This makes (3.8) evident.

Here is another key result.

Theorem 3.5. For $f_1, f_2 \in F(M)$,

$$\{\dot{f}_1, \dot{f}_2\} = \widehat{\{f_1, f_2\}} \tag{3.9}$$

Proof.

$$\{\dot{f}_1, \dot{f}_2\} = -A_{\dot{f}_1}(\dot{f}_2)$$

$$= -\widehat{A_{f_1}}(\dot{f}_2)$$

$$= -\widehat{A_{f_1}(f_2)}$$

$$= \widehat{\{f_1, f_2\}} .$$

CONSERVATION LAWS

4. A GENERAL PROLONGATION PROCESS FOR DIFFERENTIAL FORMS

Everything we have done in Sections 2 and 3 followed from formula (2.3). We can now show that this formula is a special case of a more general situation.

Let M be a manifold, $T(M)$ its tangent bundle. We have seen that

$$f \to \dot{f}$$

defines an R-linear mapping

$$F(M) \to F(T(M)) \quad.$$

This can now be extended to a mapping on differential forms:

$$\omega \to \dot{\omega}$$

for $\omega \in F^r(M)$.

To do this, work as follows. If

$$\omega = f_1 df_2 \wedge \cdots \wedge df_{r+i} \quad,$$

set

$$\dot{\omega} = \dot{f}_1 df_2 \wedge \cdots \wedge df_{r+1} + f_1 d\dot{f}_2 \wedge \cdots \wedge df_{r+1}$$
$$+ \cdots + f_1 df_2 \wedge \cdots \wedge d\dot{f}_{r+1} \qquad (4.1)$$

<u>Problem</u>. Show that this formula defines a genuine R-linear map $F^r(M) \to F^r(T(M))$ (in fact, a "first-order linear differential operator).

The following formula follows readily from (4.1):

$$\dot{\widehat{d\omega}} = d\dot{\omega} \qquad (4.2)$$

$$\widehat{\dot{\omega_1 \wedge \omega_2}} = \dot{\omega}_1 \wedge \omega_2 + \omega_1 \wedge \dot{\omega}_2 \qquad (4.3)$$

<u>Remark</u>. I do not recall seeing this operation described explicitly before. It plays a key role in studying "higher order" properties of differential equations. It can be extended to the "higher order tangent spaces" and to the "mapping element" spaces.

5. A GENERAL SETTING FOR THE THEORY OF CONSERVATION LAWS

Let X and Z be manifolds. (We now revert to the notation used in earlier chapters.) $M(X,Z)$ denotes the space of (C^∞) maps $X \to Z$. Denote a typical element of $M(X,Z)$ by $\underline{z}$. Let $M^r(X,Z)$ denote the space of r-th order <u>mapping elements</u> (i.e., "jets", in the sense of Ehresmann). Let

$$\partial^r \underline{z} \in M(X, M^r(X,Z))$$

denote the "prolongation" of $\underline{z}$.

Let

$$DE \subset M^r(X,Z)$$

be a submanifold of $M^r(X,Z)$; it defines a <u>differential equation</u> with X as the <u>independent</u>, Z as the <u>dependent</u>, variable

CONSERVATION LAWS

manifold. Let

$$\underline{S} = \{\underline{z} \in M(X,Z) : \partial^r \underline{z}(X) \subset DE\}$$

$\underline{S}$ is called the space of solutions of the differential equation. We shall suppose that there is an exterior differential system I on $M^r(X,Z)$ such that:

$$\underline{z} \in \underline{S} \iff (\partial^r \underline{z})^*(I) = 0$$

We say that I is associated with DE.

Now, let us suppose that $(DE)' \subset M^{r+1}(X,Z)$ is a submanifold of the $(r+1)$-st order mapping element space. Let $\underline{S}'$ be its space of solutions. Let us say that $(DE)'$ is a prolongation of DE if

$$\underline{S} = \underline{S}' .$$

Suppose that I' is an exterior differential system on $M^{r+1}(X,Z)$ associated to $(DE)'$. We say then that I' is a prolongation of I.

We say that

$$\theta \in F^n(M^r(X,Z))$$

$$(n = \dim X) ,$$

is a conservation law for the system DE if

$$d\theta \in I .$$

Thus, we need to study the relation between conservation laws for DE and its prolongation (DE)'. What we found, in the simple situation considered in previous sections, was that there was a natural differential operator

$$\theta \to \dot{\theta}$$

of differential forms on $M^r(X,Z)$ to differential forms on $M^{r+1}(X,Z)$ which was closely linked to the whole prolongation structure. This procedure should obviously generalize. I plan to consider this in Part B.

Chapter XI

BÄCKLUND TRANSFORMATIONS AS CONNECTIONS AND EXTERIOR DIFFERENTIAL SYSTEMS

1. INTRODUCTION

In this chapter I present a tentative geometric interpretation of a Bäcklund transformation. (It was suggested to me by Hugo Wahlquist.) We shall treat only the simplest situation, mainly using the straightforward framework suggested in Whitham's book [$\underline{1}$].

Here is that framework. Let (x,t) denote coordinates of R^2, to be identified physically with one-space and one-time variable. (We shall call this the one-D situation.) Let μ and μ' denote dependent variables. Denote their partial derivatives as

$$\mu_x, \mu_t, \mu'_x, \mu'_t, \ldots$$

Suppose given a <u>first order</u> system of differential equations interrelating μ and μ', of the following form:

$$\mu'_x = f(\mu, \partial\mu, \mu')$$

$$\mu'_t = g(\mu, \partial\mu, \mu')$$

(1.1)

Suppose that we are also given a pair of higher order differential equations which involve μ, μ' separately, of the following form:

$$D(\mu, \partial\mu, \partial^2\mu, \ldots) = 0 \qquad (1.2)$$

$$D'(\mu', \partial\mu', \partial^2\mu', \ldots) = 0 \qquad (1.3)$$

<u>Definition</u>. We say that the system (1.1) is a <u>Bäcklund transformation</u> between the differential equations (1.2) and (1.3) if the following condition is satisfied:

> For each solution $\mu(x,t)$ of (1.2), the system of equations (1.3) is <u>completely integrable</u>, and each of its solutions is also a solution of (1.3). (1.4)

<u>Remark</u>. Notice that this relation is <u>not necessarily symmetric</u> between D and D'. It does seem to be symmetric for the usual systems.

Several examples of such systems have been given in previous chapters. (For example, D and D' both the Sine-Gordon equation; D = Burger equation and D' = Heat equation The aim of this chapter is to <u>briefly</u> describe some relations

CONNECTIONS   257

between these concepts, the theory of connections and exterior
differential systems. I plan further work on these topics in
later volumes.

2. BÄCKLUND TRANSFORMATIONS AND EXTERIOR DIFFERENTIAL SYSTEMS

The relation to be described now is classical, developed
in the work of Clairin, E. Cartan, Goursat and others. However,
this material seems to have virtually disappeared from the
contemporary literature. The best classical source with which
I am familiar is the short monograph, "Le problème de Bäcklund",
by E. Goursat. He deals with a rather general case, whereas I
will stay, for simplicitly, with Bäcklund transformations of
the form (1.1).

Introduce a space M with variables labelled as follows:

$$x, t, \mu, \mu_x, \mu_t, \mu'$$

Introduce the following one-forms on M.

$$\theta_1 = d\mu - \mu_x dx - \mu_t dt \qquad (2.1)$$

$$\theta_2 = d\mu' - fdx - gdt , \qquad (2.2)$$

where f,g are functions on this space, occurring on the right
hand side of equations (1.1) defining Bäcklund transformation.

The "Bäcklund transformation" problem, as described informally in the last section, can then be transformed into the following problem:

> Let I be the differential ideal of forms on M generated by θ_1, θ_2. Find all two-dimensional solution submanifolds of I parameterized by x,t (i.e., on which $dx \wedge dt \neq 0$). Giving such a solution amounts to giving functions
>
> $$\mu(x,t), \quad \mu'(x,t), \quad \mu_x(x,t), \quad \mu_t(x,t)$$
>
> The annihilation of (2.1) guarantees that
>
> $$\mu_x = \frac{\partial}{\partial x}(\mu(x,t))$$
>
> $$\mu_t = \frac{\partial}{\partial t}(\mu(x,t))$$
>
> Thus, the differential equations
>
> $$D(\mu, \partial\mu, \ldots) = 0$$
>
> $$D'(\mu', \partial\mu', \ldots) = 0$$
>
> satisfied by these functions μ, μ' must be found. By the very logic of the situation, they will be the "Bäcklund transforms" of each other.

CONNECTIONS 259

Carrying out the analysis of the existence of two-dimensional solution submanifolds of I is now a matter of Cartan's Theory of Exterior Differential Systems. Goursat, in the work mentioned above, essentially carries out this analysis. It needs to be redone in more modern terms, a task I hope to do later on more systematically. Instead of carrying forward in this direction, I will proceed to discuss the situation from the point of view of connection theory.

3. INTRODUCTION OF A CONNECTION

Keep the notation of Section 2. Recall that M is the space of variables $(\mu,\mu',x,t,\mu_x,\mu_t)$. Let N be the space of variables

$$(\mu,x,t,\mu_x,\mu_t)\ .$$

Let

$$\pi\colon M \to N$$

be the Cartesian projection map. π is then a <u>fiber space projection map</u>, <u>with</u> R <u>as fiber</u>, N <u>as base</u>. (μ' is the <u>fiber coordinate</u>.)

The one-form θ_1 defined by (2.1) then is the pull-back under π of a one-form on N. We see from (2.2) that:

> θ_2 may be regarded as a connection form, defining an Ehresmann connection for the fiber space (M,N,π).

We will call this the Bäcklund connection.

We immediately see the following "geometric" interpretation for the Bäcklund "problem":

Find two-dimensional submanifolds of N with the following properties:
a) $dx \wedge dt$ is nonzero when restricted to the submanifold.
b) θ_1 is zero when restricted to this submanifold.
c) The connection is "flat" along the submanifold, i.e., parallel transport for the connection is independent of the path along the submanifold.

We can then write down the differential equations

$$D(\mu, \partial\mu, \ldots) = 0 \qquad (3.1)$$

for μ in terms of the curvature of this connection. Namely, a surface

$$(x,t) \to (x,t,\mu(x,t),\partial_x\mu,\partial_t\mu)$$

in N is a solution if and only if it is a solution submanifold of the exterior differential system on N generated by θ_1 and the curvature two-forms of the connection determined by θ_2.

Remark. These remarks bring out very strongly the relation between this way of looking at Bäcklund transformations and the "pseudo potentials" of Wahlquist and Estabrook. In fact, it was Wahlquist himself who first pointed out this relation to me.

4. BÄCKLUND TRANSFORMATIONS DETERMINED BY $SL(2,R)$-CONNECTIONS

We know that many of the interesting nonlinear wave equations are generated by "quadratic" pseudo potentials, in the terminology of Corones, Estabrook and Wahlquist. As we have seen in earlier chapters, they may be described by $SL(2,R)$-connections. Having just brought out the analogy between "pseudopotentials" and "Bäcklund transformations", it is natural to look for the Bäcklund transformations determined by $SL(2,R)$-connections.

Keep the notation of Section 3. M is the space of variables

$$(\mu,\mu',\mu_x,\mu_t,x,t) \quad .$$

N is the space of variables

$$(\mu, \mu_x, \mu_t, x, t) \ .$$

$\pi: M \to N$ is the projection map:

$$\pi(\mu', \mu, \mu_x, \mu_t, x, t) = (\mu, \mu_x, \mu_t, x, t) \ .$$

$$\theta_1 = d\mu - \mu_x dt - \mu_t dt \qquad (4.1)$$

is a one-form on N. Let θ_2 be the one-form on M which defines an SL(2,R)-connection for the fiber space (M,N,π). It may then be written in the following form:

$$\theta_2 = d\mu' - \omega_0 - \omega_1 \mu' - \omega_2 (\mu')^2 \ , \qquad (4.2)$$

where $\omega_0, \omega_1, \omega_2$ are one-forms <u>on</u> N, i.e., on the <u>base</u> of the fiber space. We can then compute the curvature two-forms $\Omega_0, \Omega_1, \Omega_2$ in the usual way:

$$\begin{aligned} \Omega_0 &= d\omega_0 - \omega_0 \wedge \omega_1 \\ \Omega_1 &= d\omega_1 + 2\omega_0 \wedge \omega_2 \\ \Omega_2 &= d\omega_2 + \omega_1 \wedge \omega_2 \end{aligned} \qquad (4.3)$$

We can then sum up as follows:

<u>Theorem 4.1</u>. Let a Bäcklund transformation be defined as described above via an SL(2,R) connection. Then, the partia

differential equation for $\mu(x,t)$ is determined as the solution submanifold of the exterior differential system on N generated by $\theta_1, \Omega_0, \Omega_1, \Omega_2$.

Example 1. The Burger Equation

The Bäcklund formulas are:

$$\partial_x \mu' = a\mu\mu'$$
$$\partial_t \mu' = b\mu'\partial_x(\mu) + c\mu^2\mu' \qquad (4.4)$$

(a,b,c are constants. See Whitham, [1], page 610.)
We have:

$$\theta_2 = d\mu' - (a\mu\mu')dx - (b\mu'\mu_x + c\mu^2\mu')dt .$$

We can read off the connection forms:

$$\omega_1 = a\mu dx + (b\mu_x + c\mu^2)dt$$
$$\omega_0 = 0 = \omega_2 \qquad (4.5)$$

(This means that the connection is really a G-connection, where G is a one-parameter subgroup of $SL(2,R)$.) Hence,

$$\Omega_0 = 0 = \Omega_2 .$$

$$\Omega_1 = d\omega_1$$
$$= ad\mu \wedge dx + bd\mu_x \wedge dt + 2c\mu d\mu \wedge dt \qquad (4.6)$$

In (4.6) we can replace (mod θ_1) $d\mu \wedge dx$ and $d\mu \wedge dt$ with $\mu_t dt \wedge dx$ and $\mu_x dx \wedge dt$, resulting in the exterior system generated by the following forms:

$$\begin{array}{|l|} \hline \theta_1; \\ (-a\mu_t + 2c\mu\mu_x)dx \wedge dt \\ +bd\mu_x \wedge dt \\ \hline \end{array} \qquad (4.7)$$

We see that its solution submanifolds are just the solutions of the Burger equation:

$$b\mu_{xx} + 2c\mu\mu_x - a\mu_t = 0 \qquad (4.8)$$

Example 2. The Sine-Gordon Equation

The Bäcklund equations are:

$$\begin{array}{|l|} \hline \partial_x \mu' = a \sin(\mu-\mu') - \partial_x \mu \\ \\ \partial_t \mu' = b \sin(\mu+\mu') + \partial_t \mu \\ \hline \end{array} \qquad (4.9)$$

The connection form θ_2 is then:

CONNECTIONS 265

$$\theta_2 = d\mu' - (\sin(\mu-\mu') - \mu_x)dx - (\sin(\mu+\mu') + \partial\mu_t)dt$$
(4.10)
$$= d\mu' - \omega_0 - \omega_1 \sin \mu' - \omega_2 \cos \mu'$$

where $\omega_0, \omega_1, \omega_2$ are one-forms on the variables $(x, t, \mu_x, \mu_t, \mu)$. This is a connection form associated to the Lie algebra of vector fields

$$\frac{\partial}{\partial\mu'}, \quad \sin(\mu')\frac{\partial}{\partial\mu'}, \quad \cos(\mu')\frac{\partial}{\partial\mu'},$$

which is again just $SL(2,R)$ acting on R.

5. BÄCKLUND TRANSFORMATIONS IN TERMS OF CONNECTIONS WITH TWO-DIMENSIONAL FIBERS

So far in this chapter I have developed the connection theory of Bäcklund transformations in close analogy with the theory of Estabrkkl-Wahlquist pseudo potentials. Here is a slightly different appraoch.

Start off again with the following equations:

$$\mu'_x = f(\mu, \mu_x, \mu_t, \mu')$$

$$\mu'_t = g(\mu, \mu_x, \mu_t, \mu') .$$

As before, convert this into the problem of finding two-dimensional solution submanifolds of the exterior differential system generated by the following one-forms:

$$\theta_1 = d\mu - \mu_x dx - \mu_t dt \qquad (5.1)$$

$$\theta_2 = d\mu' - f dx - g dt$$

Construct a fiber space (E,B,ϕ) in a slightly different way from the one (M,N,π) constructed in previous sections. Namely, let:

E = space of variables $(x,t,\mu,\mu',\mu_x,\mu_t)$

B = space of variables (x,t,μ_x,μ_t)

$\phi(x,t,\mu,\mu',\mu_x,\mu_t) = (x,t,\mu_x,\mu_t)$.

This is a fiber space with a <u>two-dimensional fiber</u>. The forms (5.1) determine a <u>connection</u> for this fiber space. Its "group" is of course just essentially the "trivial extension" of the group for the connection with a one-dimensional fiber described in Section 4. One could readily investigate the relation between its curvature and the differential equations which are linked via the Bäcklund transformation, but I will not go into this here. The esthetic virtue of this approach is that it seems to treat the variables μ and μ' in a more symmetric manner.

6. ANOTHER TWO-VARIABLE CONNECTION APPROACH TO THE SINE-GORDON BÄCKLUND

Consider equations (4.9). Set:

$$y = \mu - \mu' \tag{6.1}$$

$$z = \mu' + \mu \tag{6.2}$$

Equations (4.9) become:

$$\partial_x z = a \sin y$$
$$\partial_t y = -b \sin z \tag{6.3}$$

Write it in the form of an exterior system:

$$\theta_1 \equiv dz - a \sin y \, dx - z_t \, dt$$
$$\theta_2 = dy - y_x \, dx + b \sin z \, dt \tag{6.4}$$

Let E be the space of variables

$$(y, z, y_x, z_t, x, t)$$

B the space of variables (x, t, y_x, z_t). $\pi: E \to B$ is defined as the <u>Cartesian projection</u>. (y,z) are the fiber coordinates. We see that equations θ_1, θ_2 define an $SL(2,R) \times SL(2,R)$ <u>connection</u> for this bundle.

Let us complute the <u>curvature form</u> for this connection using the technique described in IM, Vol. 10:

$$d\theta_1 = -a \cos y \, dy \wedge dx - dz_t \wedge dt$$

$$= -a \cos y (y_x dx + b \sin z \, dt) \wedge dx - dz_t \wedge dt + \cdots$$

(The term ... are in the Grassman ideal generated by θ_1, θ_2.)

$$= -ab \cos y \sin z \, dt \wedge dx - dz_t \wedge dt \quad .$$

Similarly,

$$d\theta_2 = -dy_x \wedge dt + b \cos z \, dz \wedge dt$$

$$= -dy_x \wedge dx + b \cos z (a \sin y \, dx) \wedge dt + \cdots$$

Set:

$$\Omega_1 = +ab \cos y \sin z \, dx \wedge dt - dz_t \wedge dt$$

$$\Omega_2 = -dy_x \wedge dx + ab \cos z \sin y \, dx \wedge dt$$

They are the <u>curvature forms</u> of this connection. Let us look for the <u>two-dimensional submanifolds of</u> B <u>on which</u> $dx \wedge dt \neq 0$ and for which <u>annihilate the curvature forms</u>. Thus, we look for z, y as functions $y(x,t)$, $z(x,t)$. Also, <u>impose the condition</u>:

$$z_t = \frac{\partial z}{\partial t}$$

$$y_x = \frac{\partial y}{\partial x}$$

(I do not see any way these conditions are implied only by the vanishing of the curvature.) Thus, the conditions that Ω_1, Ω_2 vanish on this two-dimensional submanifold (i.e., that the connection be flat on such a submanifold, or that parallel transport be independent of the path chosen within that submanifold) is that

$$\partial_{tx} z = ab \cos y \sin z$$

$$\partial_{tx} y = -ab \cos z \sin y$$

When one goes back to μ, μ', one sees that these equations are equivalent to Sine-Gordon.

This gives us a method--which hopefully generalizes--to derive the Sine-Gordon from a connection in a two-dimensional fiber vector space. Of course, this approach has the great virtue of treating the pair of variables (y,z) --or (μ,μ') -- in a much more symmetric manner than the approaches given earlier.

Chapter XII

SOME BRIEF REMARKS CONCERNING THE ALGEBRAIC SETTING FOR THE THEORY OF EXTERIOR DIFFERENTIAL SYSTEMS

1. INTRODUCTION

The Theory of Exterior Differential Systems is far from complete, even on a rather elementary level.

There are a whole complex of problems--some as difficult as any in mathematics (or physics), some only "elementary" in a technical sense, but probably requiring computational skills going far beyond the capabilities of most of today's mathematicians. (One prominent aspect of Cartan's genius was precisely this skill!) In this chapter I will present several general remarks.

Consider a manifold X and an exterior differential system, ED, on M, i.e., a differential ideal of differential forms. Given a point x ε X, denote by

$$\Lambda(X_x)$$

the Grassman <u>algebra</u> of the differential forms at p. Let

$$ED(x) \subset \Lambda(X_x)$$

be the <u>subalgebra</u> (under exterior product) resulting from evaluating all of the forms ED at x.

Let $GL(X_x)$ be the group of linear automorphisms of the tangent space X_x. It extends to act as a group of algebraic automorphisms on $\Lambda(X_x)$. The "algebraic" problem of obvious relevance to the theory of exterior differential systems is to "study" (and possibly "parameterize") the <u>orbits</u> of $GL(X_x)$ and to study the subgroup which maps $ED(x)$ into itself.

In general, this is obviously a matter of "invariant theory", of a very difficult kind. (For example, the action of the general linear groups on third degree Grassman elements has apparently never been adequately studied from this point of view.) However, scattered through Cartan's works are many special calculations, and I plan to use these as a clue to begin such a systematic study.

Of course, related to this (but much harder!) is the usual "equivalence problem": Given ED, ED', when can one find a diffeomorphism

$$\phi: X \to X'$$

such that

$$\phi^*(ED') = ED \quad ?$$

In simple cases, Cartan often solved this by the classical trick of finding "canonical forms". I will also present a few semi-trivial remarks here about systems to which it readily applies.

ALGEBRAIC SETTING

2. SYSTEMS GENERATED ALGEBRAICALLY BY ZERO- AND ONE-FORMS

Let ED denote a differential ideal which is <u>generated as an algebra</u> by zero and one-forms. (It is then, of course, "completely integrable", and the "Frobenius theorem" applies, at least in the nonsingular situation.) Of course, the "zero-forms" are <u>functions</u> on X. They can be set equal to zero, and X can be replaced by this subset. In the "generic" situation (which we assume) this subset will be a <u>submanifold</u>, hence we can restrict attention to the <u>case where</u> ED <u>is generated algebraically by one-forms</u>.

The algebraic invariants are now obvious. Let $(ED)^1$ denote the subset of ED consisting of forms of degree one. (It is an $F(X)$-submodule of $F^1(X)$.) These values of these forms at x define a linear subspace of X_x^d, the vector space of one-covectors at x. Let us denote this subspace as

$$(ED)^1_x .$$

Let

$$m = \dim (ED)^1_x .$$

Assuming that m is constant as x ranges over X (again, this is the "generic" situation), it defines the obvious (and the only) <u>invariant</u> of the system, since $GL(X_x)$ acts transitively on the space of such linear subspaces.

The local "canonical form" is also determined by m. The Frobenius complete integrability theorem implies the existence (locally) of functionally independent functions $f_1,\ldots,f_m$ such that

$$df_1,\ldots,df_m$$

generates ED. Two such systems with the same m are also "equivalent", in the sense that there is (locally) a diffeomorphism sending one to the other.

Of course, the global equivalence of two such systems is a more complicated and difficult story, and is now in full development under the name: The Theory of Foliations.

3. SYSTEMS GENERATED ALGEBRAICALLY BY ONE-FORMS AND A SINGLE TWO-FORM

This is the next most complicated situation. Again, let us suppose that zero-forms have been eliminated by restriction to submanifolds. We assume then that ED is generated algebraically by one-forms $\theta^1, \theta^2, \ldots$ and a single two-form Since ED is a differential ideal, we have relations of the form

$$d\theta^1 = f^1 \omega \qquad (3.1)$$

$$d\theta^2 = f^2 \omega$$

$$\vdots$$

$$d\omega = 0 \qquad (3.2)$$

ALGEBRAIC SETTING

Luckily, the linear algebra needed to find the orbits of $GL(X_x)$ is well known. Let

$$\gamma_x \subset X_x$$

be the linear subspace of X_x consisting of the vector $v \in X_x$ such that

$$\theta^i(v) = 0 \qquad (3.3)$$

Let

$$\gamma_x^{cc} = \{v \in \gamma_x : \omega(v,\gamma_x) = 0\}$$

("cc" stands for Cauchy characteristics.) Set:

$$m = \dim \gamma_x \qquad (3.4)$$

$$m' = \dim \gamma_x^{cc} \qquad (3.5)$$

The invariants under $GL(X_x)$ are just (m, m').

As x_1 varies, the assignment

$$x \to \gamma_x^{cc}$$

defines a vector field system on X which is <u>completely integrable</u>. (An "involutive distribution", in Chevalley's language.) Suppose that

$$\pi: X \to Y$$

is a quotient space for this foliation. The classic properties of "Cauchy characteristics" now imply that there is an exterior system (ED)' in Y such that:

> ED is generated by $\pi^*((ED)')$

> (ED)' has no Cauchy characteristics

Here is what is involved in terms of local coordinates. Let (x^i, y_i) be "canonical coordinates" for ω, i.e., a set of functionally independent functions such that:

$$\omega = dy_i \wedge dx^i . \qquad (3.6)$$

(It is "Darboux's theorem" that such coordinates exist locally. See DGCV.) Let us take into account relations (3.1). Apply to both sides:

$$df^1 \wedge (dy_i \wedge dx^i) = 0 \qquad (3.7)$$

These imply that:

> The functions $f^1, f^2, \ldots$ are functions of (x,y) alone. $\qquad (3.8)$

ALGEBRAIC SETTING

Thus, we can find one-forms η^1, η^2 in <u>terms of the variables</u> (x,y) <u>alone</u> such that

$$d\eta^1 = f^1 \omega$$
$$d\eta^2 = f^2 \omega \qquad (3.9)$$
$$\vdots$$

In particular, we have:

$$d(\eta^1 - \theta^1) = 0$$
$$d(\eta^2 - \theta^2) = 0$$
$$\vdots$$

Hence, there are functions $z^1, z^2, \ldots$ on X such that:

$$\theta^1 = dz^1 - \eta^1$$
$$\theta^2 = dz^2 - \eta^2 \qquad (3.10)$$
$$\vdots$$

Now, let us proceed further to study the η's. If all the $f^1, f^2, \ldots$ are zero, then

$$\theta^1 = dz^1$$
$$\theta^2 = dz^2$$
$$\vdots$$

is the canonical form. Putting this to the side (and the "non-generic case" where all the f's vanish at one point, but not identically), we can suppose that:

$$d\theta^1 = \omega \ .$$

Hence, we can chose z^1 so that:

$$\theta^1 = dz^1 - y_i dx^i \qquad (3.11)$$

Formulas (3.10) and (3.11) then define a "partial canonical form".

Remark. The most complete treatment of this situation in the classical literature with which I am familiar is in E. Cartan's article, "Sur certaines expressions différentielles et le problème de Pfaff", Collected Works, Part II, Vol. 1, or in Goursat's book, "Le problème de Pfaff". The reader is strongly encouraged to look at these two works for himself.

Here are some examples.

Example 1. ED <u>is generated by a single one-form</u>

The analysis given above implies that the canonical form is:

$$\theta = dz - y_i dx^i \ .$$

ALGEBRAIC SETTING

Example 2. ED is generated by two one-forms

The partial canonical form is:

$$\theta^1 = dz^1 - y_i dx^i$$

$$\theta^2 = dz^2 + \eta^2 \quad , \tag{3.12}$$

where η^2 is a one-form in the variables (x,y) alone, z^1, z^2 are functions on X, $d\eta^2 = d\theta^2 = f^2 d\theta^1$. Hence,

$$d(f^2) \wedge d\theta^1 = 0 \tag{3.13}$$

Exercise. If the dimension of (x,y) space is greater or equal to 4, show that relation (3.13) forces

$$df^2 = 0 \quad . \tag{3.14}$$

Using relation (3.14),

$$d(\eta^2 - f^2 \theta^1) = 0 \quad ,$$

or

$$\eta^2 - f^2 \theta^1 = d\omega^2 \quad .$$

Hence, the canonical form for the system is just (in case dimension (x,y) space is ≥ 4):

$$\boxed{\begin{aligned} \theta_1 &= dz^1 - y_i dx^i \\ \theta_2 &= dz^2 + cy_i dx^i \end{aligned}} \tag{3.15}$$

where z_1, z_2 are "arbitrary functions", c is a <u>constant</u>.

As an important qualitative remark, notice that solution manifolds for ED can be written down by "formulas" (i.e., by differentiation) once the system is in its canonical form ED. Throwing the system into this canonical form just involves operating within a C^∞ context. Hence, the solution manifold of ED can be found by means of C^∞ <u>operations</u>. Thus, in a sense, these "canonical form" arguments give an <u>algorithm</u> for finding (locally!) the solution submanifolds. They may be thought of as a sort of <u>Galois Theory</u> for the system.

Here is a classical system to which this can be applied.

<u>Example 3</u>. <u>First order P.D.E.'s for one unknown function</u>

Suppose

$$\begin{aligned} F_1(x, \partial_x \mu, \mu) &= 0 \\ F_2(x, \partial_x \mu, \mu) &= 0 \\ &\vdots \end{aligned} \tag{3.1}$$

is such a system. Introduce:

ALGEBRAIC SETTING

$$\theta = d\mu - y_i dx^i$$

X = space of (μ, μ_x, x), <u>constrained by relations</u> (3.16).
ED = system generated by θ <u>restricted to</u> X.

Many of the details and examples needed to understand this situation are covered in the work by Cartan and Goursat which was mentioned above, to which the reader should refer.

Let us compare this set-up with that obtained for a single second order partial differential equation in two independent variables, e.g., one of the form:

$$\mu_{tt} = F(x, t, \mu, \mu_x, \mu_t, \mu_{xt}) \quad .$$

Set:

$$\theta_1 = d\mu - \mu_t dt - \mu_x dx$$

$$\theta_2 = d\mu_t - F dt - \mu_{xt} dx$$

We see that the basic algebraic problem here involves study of a pair $d\theta_1$, $d\theta_2$ of two-covectors at points of x. We now turn to this problem.

4. ALGEBRAIC STUDY OF PAIRS OF SKEW SYMMETRIC BILINEAR FORMS BY MEANS OF KRONECKER'S THEORY OF PENCILS OF MATRICES

Forget about manifold theory for this section and study a vector space V of dimension n, with an arbitrary scalar field K. Let

$$\omega_1, \omega_2 : V \times V \to K$$

be skew-symmetric, bilinear forms on V. Let λ be a variable in K, and set:

$$\omega(\lambda) = \omega_1 + \lambda \omega_2 . \qquad (4.1)$$

$\omega(\lambda)$ is a *pencil of forms*. Let

$$V[\omega] = \{v_0 + \lambda v_1 + \lambda^2 v_2 + \cdots : v_0, v_1, \ldots \in V\}$$

be the space of polynomials in λ, with coefficients in V. $\omega(\lambda)$ defines a skew-symmetric, bilinear map

$$V[\lambda] \times V[\lambda] \quad K[\lambda]$$

in the obvious way. $V[\lambda]$ is a *free module over the polynomial ring* $K[\lambda]$.

In Gantmacher's "Theory of Matrices", Chapter 12, Section 6 it is shown how $V[\lambda]$ splits up relative to $\omega(\lambda)$. First, V is a direct sum of linear subspaces

$$V = V'' \oplus V'' .$$

$$V[\lambda] = V'[\lambda] \oplus V''[\lambda] .$$

$$\omega(\lambda)(V'[\lambda], V''[\lambda]) = 0$$

$\omega(\lambda)$ restricted to $V'[\lambda]$ is *regular*, i.e., its determinant

ALGEBRAIC SETTING 283

does not vanish <u>identically</u> in λ. $\omega(\lambda)$ restricted to $V''[\lambda]$ is <u>singular</u>, in the sense that its determinant does vanish identically.

Isomorphism of two pairs (ω_1, ω_2), (ω_1', ω_2') now depends on the isomorphism of the pencils $\omega(\lambda)$, $\omega'(\lambda)$, which in turn depends (by the basic Kronecker theory described by Gantmacher) on the isomorphism of the singular and regular parts. In turn, they can be decomposed, as described by Gantmacher (supplemented, perhaps, by IM, Vol. 9, Chapter 8) into irreducible and orthogonal pairs.

5. A GENERAL INVARIANT-THEORETIC SETTING

Of course, the Kronecker pencil theory is but one method for the description of "invariants" of certain Lie group actions. Here is a brief description of a very general setting.

Again, let V be a finite dimensional vector over a field K. Let V^d denote the dual space of V.

$$V^d \wedge V^d$$

denotes the space of <u>skew-symmetric bilinear forms</u> on V. $GL(V)$, the group of linear automorphisms of V of course acts on $V^d \wedge V^d$. It is, of course, this action that we are most concerned with.

A polynomial map

$$f: V^d \wedge V^d \to K$$

is called a <u>relative invariant of weight</u> n if

$$f(\alpha(\omega)) = (\det \alpha)^n f(\omega)$$

for $\alpha \in GL(V)$.

(For example, $f(\omega)$ could be the determinant of the matrix of ω with respect to a basis of V. More invariantly, if $\dim V = 2m$ and if ω_0 is a nonzero element of

$$\underbrace{V^d \wedge \cdots \wedge V^d}_{2m \text{ times}}$$

then

$$\omega_0 f(\omega) = \underbrace{\omega \wedge \cdots \wedge \omega}_{m \text{ times}}$$

Now, suppose

$$(\omega_1, \ldots, \omega_r)$$

are a collection of two-forms on V. An important problem of Cartan's theory is the description of the "invariants" (e.g., perhaps the "parameterization" of the orbits) of the action of $GL(V)$ on such r-tuples. Here is one approach.

ALGEBRAIC SETTING

Let $\omega \to f(\omega)$ be such a relative invariant. Form

$$(\lambda_1,\ldots,\lambda_r) \to \lambda_1\omega_1 + \cdots + \lambda_r\omega_r \;,$$

where $\lambda_1,\ldots,\lambda_r$ are variables in K. Then, form

$$f(\lambda_1\omega_1 + \cdots + \lambda_r\omega_r) \;,$$

a polynomial in $\lambda_1,\ldots,\lambda_r$. The coefficients of this polynomial are obviously relative invariants of the action of GL(V) on such r-tuples of forms. In particular, <u>singular orbits</u> are typically obtained by setting such relative invariants equal to zero. This is well known in the classical literature and has recently been revived in a very fancy way in the modern algebro-geometric literature.

6. ALGEBRAIC INVARIANTS OF THE SECOND ORDER PARTIAL DIFFERENTIAL EQUATION IN TIME INDEPENDENT VARIABLES

Here is a traditional example

$$f(\mu, \partial\mu, \partial^2\mu) = 0 \qquad (6.1)$$

Introduce variables

$$y, t, \mu, \mu_y, \mu_{yy}, \mu_{tt}, \mu_{yt} \;.$$

X is the submanifold of R^8 obtained by setting $f = 0$

$$\theta_1 = d\mu - \mu_y dy - \mu_t dt$$

$$\theta_2 = d\mu_y - \mu_{yy}dy - \mu_{tt}dt$$

$$\theta_3 = d\mu_t - dt - \mu_{yt}dx$$

Given a point $x \in X \subset R^8$ ($= M^2(R^2,R)$, in terms of the "mapping element" notations used in previous chapters), set

γ_x = linear subspace of X_x determined by setting $\theta_1 = 0 = \theta_2 = \theta_3$.

γ_x is then a four-dimensional vector space. Let $\omega_1, \omega_2, \omega_3$ be the two-forms $d\theta_1, d\theta_2, d\theta_3$ restricted to γ_x. Introduce three new variables

$$\lambda = (\lambda_1, \lambda_2, \lambda_3)$$

$$\omega(\lambda) = \lambda_1\omega_1 + \lambda_2\omega_2 + \lambda_3\omega_3$$

$$f(\lambda) = \omega(\lambda) \wedge \omega(\lambda)$$

The <u>singular values of</u> λ are those for which

$$f(\lambda) = 0 .$$

These can also be characterized as the values of λ for which:

$$(\lambda_1 d\theta_1 + \lambda_2 d\theta_2 + \lambda_3 d\theta_3) \wedge \theta_1 \wedge \theta_2 \wedge \theta_3 \wedge df = 0$$

When one makes this condition explicit (see Goursat's "Lecons sur le Problème de Pfaff", page 286), one sees the main local properties of the partial differential equation

$$f(\mu, \partial\mu, \partial^2\mu) = 0$$

exhibited in a remarkably simple way. (I owe this observation to a lecture by Professor Robert Gardner at the 1976 Ames (NASA) Institute/Seminar on Differential and Algebraic Geometry for the Control Engineer.)

I believe that many other of the examples treated by Cartan and his contemporaries can be unified from this point of view--parameterization of the orbits of GL(V) acting on the skew-symmetric tensor spaces over a vector space V. This will be a main topic of a later volume.

Chapter XIII

DEFORMATIONS OF EXTERIOR DIFFERENTIAL SYSTEMS AND SINGULAR PERTURBATION THEORY

1. INTRODUCTION

The topic called "singular perturbation theory" is one of the most active and important in contemporary mathematical physics and engineering. The traditional setting may be described as follows:

Consider a second order ordinary differential equation

$$\epsilon \frac{d^2 x}{dt^2} = f\left(x, \frac{dx}{dt}, t\right) \qquad (1.1)$$

ϵ is a parameter. As $\epsilon \to 0$, the equation changes "character". One might ask how the solutions for $\epsilon > 0$ and for $\epsilon = 0$ are related. One is often interested in studying how the solutions satisfying given initial or boundary conditions behave. Notice that the system (1.1) changes order for $\epsilon = 0$, hence a smaller number of such conditions may be imposed. This reflects the "singular" nature of the perturbation problem.

To see the relation with the theory of exterior differential systems, let us convert the equation (1.1) into such a system. Introduce variables

$$(x, \dot{x}, t) \quad ,$$

and the following one-forms:

$$\theta_1 = dx - \dot{x}\, dt$$

(1.2)

$$\theta_2 = \epsilon\, d\dot{x} - f\, dt$$

Let $(ED)_\epsilon$ denote the differential ideal of forms generated by forms θ_1, θ_2. We obtain in this way a one-parameter family

$$\epsilon \to (ED)_\epsilon$$

of ideals of forms which varies smoothly with ϵ. In contemporary mathematics, such objects are frequently encountered, and are given the generic name "deformations". Although I do not recall having seen a discussion of this particular type of "deformation", one familiar with the other areas can carry over many of the qualitative ideas. (Much of the original work was by K. Kodaira and D.C. Spencer, in connection with deformations of complex manifolds, and certain other geometric structures. I wrote a series of papers, "Analytic Continuation of Group Representations", adapting some of these ideas to physics and Lie group representation theory. In fact, I believe there are especially close links between this work and the topic to be treated here.)

Return to system (1.2). At $\epsilon = 0$, it goes over "smoothly" to the following one:

SINGULAR PERTURBATIONS

$$\theta_1^0 = dx - \dot{x}\, dt$$
$$\theta_2^0 = f\, dt \quad .$$
(1.3)

This is in a sense a "reducible" system, analogous to the notion of a "reducible algebraic set" in algebraic geometry. In particular, its one-dimensional solution manifolds are curves of the following type:

$$\boxed{\begin{array}{c} \dot{x} = \dfrac{dx}{dt} \\[4pt] t = t \\[4pt] f\left(x, \dfrac{dx}{dt}, t\right) = 0 \end{array}}$$
(1.4)

or

$$\boxed{\begin{array}{c} t = \text{constant} \\ x = \text{constant} \end{array}}$$
(1.5)

Thus, we would expect on geometric grounds that certain types of solution manifolds of (1.2) go over (as $\epsilon \to 0$) to solutions of (1.4), others to solutions of (1.5). (This is an elementary, but probably profound, "geometric" explanation of what has happened to the "missing" boundary or initial conditions as $\epsilon \to 0$. They "reappear" among the "degenerate" solutions of type (1.5). In fact, one of Sophus Lie's great contributions to the geometric study of differential equations was precisely this type of geometric extension of the meaning of "solution" of a differential equation.)

Remark. This phenomenon--"irreducible" algebraic sets deforming to " reducible" ones--is quite common in algebraic geometry. For example, for each $\epsilon \neq 0$, the set of $(x,y) \in \mathbb{C}^2$ such that

$$x^2 - y^2 = \epsilon^2$$

is irreducible. As $\epsilon \to 0$, it goes over into

$$x^2 - y^2 = 0 \quad,$$

which is, of course, reducible.

We shall now consider the simplest case where everything can be understood explicitly--the constant coefficient linear differential equation--and then go on to present some general remarks in terms of the theory of exterior differential systems

… SINGULAR PERTURBATIONS

2. SINGULAR PERTURBATION OF SECOND ORDER, LINEAR, CONSTANT COEFFICIENT ORDINARY DIFFERENTIAL EQUATIONS

Consider the following equation

$$\epsilon \frac{d^2 x}{dt^2} + a \frac{dx}{dt} + bx = 0 \qquad (2.1)$$

Its solutions are, of course

$$x(t) = c_+ e^{\lambda_+ t} + c_- e^{\lambda_- t}, \qquad (2.2)$$

with

$$\lambda_\pm = \frac{-a \pm \sqrt{a^2 - 4\epsilon b}}{2\epsilon} \qquad (2.3)$$

In terms of $(x, \dot{x}, t)$-space, the following curves are constructed

$$S \to \begin{cases} x(t) = c_+ e^{\lambda_+ t(s)} + c_- e^{\lambda_- t(s)} \\ \dot{x} = \left(\lambda_+ c_+ e^{\lambda_+ t(s)} + \lambda_- c_- e^{\lambda_- t(s)} \right) \\ t = t(s) \end{cases} \qquad (2.4)$$

c_+, c_- are functions of ϵ. Now,

$$\lambda_- \to \infty \quad \text{as} \quad \epsilon \to 0$$

$$\lambda_+ \to -\frac{b}{a} \quad \text{as} \quad \epsilon \to 0$$

We can then construct a one-parameter family of solutions as follows:

$$s \to \begin{cases} x(s) = c_+ e^{\lambda_+ s} \\ \dot{x} = \lambda_+ c_+ e^{\lambda_+ s} \\ t = s \end{cases} \quad (2.5)$$

This family of solutions goes smoothly, as $\epsilon \to 0$, into solutions of the "unperturbed" equation:

$$a \frac{dx}{dt} + bx = 0$$

However, there are other families of solution curves which have smooth limits as $\epsilon \to 0$. For example,

$$s \to \begin{cases} t = \epsilon s \\ \dot{x}(s) = \lambda_-(c\epsilon) e^{\lambda_- \epsilon s} \\ x(s) = \epsilon c e^{\lambda_- \epsilon s} \end{cases}$$

c is a constant independent of ϵ. As $\epsilon \to 0$, this goes into the following curve:

$$s \to \begin{cases} t = 0 \\ \dot{x}(s) = c e^{-as} \\ x(s) = c \end{cases}$$

SINGULAR PERTURBATIONS

Thus, we encounter in a very concrete way the general phenomenon guessed at at the end of the previous section--the system obtained as $\epsilon \to 0$ is "reducible", in the sense that it consists of two connected families, each of which is the solution manifold set of a different exterior system, namely:

$$dx - \frac{b}{a} x \, dt = 0$$

$$\dot{x} = \frac{b}{a} x$$

and

$$dx = 0 = dt \quad ; \quad \dot{x} \text{ arbitrary} .$$

Exercise. Determine precisely which curves of the two-parameter family (2.4) have a smooth limit as $\epsilon \to 0$.

3. SINGULAR PERTURBATION VIA THE RICATTI EQUATION

Here is another simple calculational example which I believe has some merit as a guide to general ideas. Consider again the differential equation of the previous section:

$$\epsilon \frac{d^2 x}{dt^2} + a \frac{dx}{dt} + bx = 0 \tag{3.1}$$

Convert it into a <u>first order Ricatti equation</u> by the well known procedure:

$$y = \frac{dx}{dt} / x \qquad (3.2)$$

$$\frac{dy}{dt} = \frac{d^2x}{dt^2} / x - \left(\frac{dx}{dt}\right)^2 / x^2$$

$$= \left(-\frac{a}{\epsilon} \frac{dx}{dt} - \frac{b}{\epsilon} x\right) / x - y^2$$

$$= -\frac{a}{\epsilon} y - \frac{b}{\epsilon} - y^2 \qquad (3.3)$$

Equation (3.3) is a <u>Ricatti equation</u>. To "guess" its singular perturbation behavior as $\epsilon \to 0$, convert it into an exterior system:

$$\epsilon \, dy + (ay + b + \epsilon y^2) \, dt = 0 \qquad (3.4)$$

Now, let $\to 0$, obtaining the system:

$$(ay + b) \, dt = 0 \quad . \qquad (3.5)$$

Thus, either

$$dt = 0 \quad ,$$

or

$$ay + b = 0 \quad ,$$

or

SINGULAR PERTURBATIONS

$$y = -\frac{b}{a} \tag{3.6}$$

Regard (3.2) as an exterior system:

$$dx - xy\, dt = 0. \tag{3.7}$$

A curve for which $dt = 0$, then--if it satisfies (3.7)--also satisfies

$$dx = 0.$$

This is one type of "generalized solution" (say, in the sense of Lie) of the equation obtained by letting $\epsilon \to 0$ in (3.1).

Consider also the solution of (3.5) for which $dt \neq 0$. It then satisfies (3.6), hence solving (3.7) as a differential equation, we obtain:

$$x = ce^{-(b/a)t}$$

This is the other type of solution for the system at $\epsilon = 0$ that we found in Section 2.

We can also prove <u>directly</u> that there are solutions of (3.3) which go over to the solution (3.6). The tip-off to this is Equation (2.5), giving a one-parameter family of solutions of the second order equation which behave "correctly" as $\epsilon \to 0$. For this family, notice that

$$y = \frac{\dot{x}}{x} \equiv \underline{\text{constant in}}\ t.$$

We can also find this family by looking for the solutions of (3.3) <u>which are independent of</u> t, i.e., which are solutions of:

$$0 = y^2 + \frac{a}{\epsilon} y + b\epsilon^{-1},$$

or

$$y = \lambda_+ \text{ or } \lambda_-.$$

We know already (from Section 2) that only λ_- has a finite limit as $\epsilon \to 0$. When (3.2) is solved for $y \equiv \lambda_-$, this provides the one-parameter family of solutions of (2.1) which go over, as $\epsilon \to 0$, to solutions of a $\frac{dx}{dt} + bx = 0$, <u>without change in parameterization</u>.

4. SOLUTION SUBSETS OF EXTERIOR DIFFERENTIAL SYSTEMS

Let us now consider very general exterior systems, and try to formulate some general ideas which will enable us to understand how to "deform" exterior systems and their solution manifolds.

Let X be a <u>real analytic</u> manifold. Let ϵ be a real parameter, varying in a small interval about $\epsilon = 0$. An <u>exterior differential system</u> for X is a differential ideal of differential forms on X. We will, for the rest of this chapter, abbreviate it to "system".

Let $\epsilon \to (ED)_\epsilon$ be a one-parameter family of systems. It is said to <u>vary smoothly with</u> ϵ if there is a finite set $\theta_1^\epsilon, \theta_2^\epsilon, \ldots$ of (real analytic) one-forms <u>which depends real-analytically on</u> ϵ which generates $(ED)_\epsilon$. Such a family is said to be a <u>deformation</u> of $(ED)_0$.

We now must formulate a suitable concept of "deformation" of submanifolds. There are, in fact, certain technical difficulties here. I will propose one way of treating them, but I do not necessarily claim that my way is optimal.

Let S be a subset of X, and let ED be an exterior system on X.

Definition. S is an m-<u>dimensional solution submanifold</u> of ED if the following conditions are satisfied:

a) For every smooth map

$\phi: Y \to X$

between manifolds such that

$\phi(Y) \subset S$,

we have:

$\phi^*(ED) = 0$.

b) For each point $x \in S$, there is an open subset $O \subset X$ containing x such that $O \cap S$ can be exhibited as an m-dimensional submanifold of X in the usual sense.

Remark. The point is that it will be necessary to think about "submanifolds" in a way that is independent of how they are parameterized, then construct a suitable "space" of such submanifolds. This is a notoriously complicated problem (for example, it appears traditionally in the calculus of variations) but no doubt there are now available adequate conceptual tools for handling it. I am not particularly an expert in these matters, so will not become involved here with the details.

Let SUB(X) denote the space of all subsets of X. Let

$$SUB(X, ED, m) \subset SUB(X)$$

denote the set of all such subsets which are solution subsets of ED. This "functorial" assignment

$$ED \to SUB(X, ED, m)$$

is analogous to the basic "functor" of algebraic geometry which assigns to each ideal in the ring of polynomials in n complex variables the algebraic subset of $\mathbb{C}^n$ consisting of the points at which all polynomials of the ideal vanish. (Recall that a basic intuitive idea underlying Cartan's theory of Exterior Differential Systems is the development of the analogy between differential equation theory and algebraic geometry!)

In order to do "deformations", we must have available suitable ideas of "continuity" and "smoothness" for curves in

SUB(X). This is not a straightforward business, and one that I do not want to cover in any detail here. Basically, it is a question of putting appropriate <u>topological</u> and <u>infinite dimensional manifold structures</u> on appropriate subsets of SUB(X). (For example, if X is a vector space, one can consider the subset of SUB(X) consisting of the <u>linear</u> subspaces, i.e., the <u>Grassman manifold</u>. It is then a familiar-- but important--story of how one goes about defining its manifold structure.)

At any rate, once we have available appropriate notions of <u>deformation</u> of exterior differential systems and topologies for subsets of SUB(X), we are in <u>position</u> to <u>ask</u> whether deformations of systems can be "followed" by deformations of the solution manifolds. We cannot go any further, because not much is known at this level of generality. I am mentioning it here because it seems to me to be the key <u>geometric</u> problem underlying singular perturbation theory. Such unifying geometric <u>insights</u> are, I believe, one of the major contributions we differential geometers can make to applied mathematics! We will call this <u>the problem of parameterization of the solution manifolds</u>.

<u>Remark</u>. Instead of varying over a real number interval, ϵ could, of course, vary over more general spaces, e.g., Euclidean spaces of higher dimension. From the point of view of

deformation theory, it is particularly important when the appropriate topology for the subsets of SUB(X) can be chosen so that the subsets are compact. (For example, in the usual topology, the Grassman manifold of linear subspaces of a vector space is compact.) In the next section we shall review several simple examples from this point of view.

Example. $a \frac{dx}{dt} = bx + c$: a, b and c are free parameters X is R^2, the space of variables (x,t).

$$\theta = a\, dx - (bx-c)\, dt$$

ED is generated by θ.

First suppose that $a \neq 0$. Set:

$$f(x,t) = bx + c - ke^{abt}.$$

((a,b,c,k) are constants.) The solution submanifolds are the subsets of R^2 determined by

$$f(x,t) = 0.$$

Exercise. Show that a manifold structure can be imposed on the space of solution submanifolds of θ, so that the assignment

$$(a,b,c,k) \to \{(x,t): f(x,t) = 0\}$$

is a real analytic map. If one also allows $a = 0$, can the solution manifolds be reasonably parameterized, e.g., so as to

SINGULAR PERTURBATIONS

become a compact manifold? Can you "identify" this manifold, e.g., as a coset space of a Lie group?

Exercise. Carry out the same program for the following differential equations:

$$a \frac{dx^2}{dt^2} + b \frac{dx}{dt} + cx + d = 0$$

$$a \frac{dx}{dt} = bx^2 + cx + d$$

$$a \frac{dx}{dt} = bx^3 + cx^2 + dx + e$$

(In this case, the "general solution" involves elliptic functions.)

Remark. Of course, in all these examples, we are dealing with completely integrable exterior systems, so that the "general solution depends on a finite number of parameters." In more typical situations, the "general solution" will depend on "arbitrary functions", and this parameterization problem--together with the corresponding deformation and singular perturbation problems--becomes a good deal more challenging and complicated.

Bibliography

1. R. Abraham and J. Marsden, Foundations of Mechanics, W.A. Benjamin, New York, 1967.

1. E. Artin, Galois Theory, University of Notre Dame, 1944.

1. E. Cartan, Les Systemes Exterieures et leurs Applications Géométriques, Hermann, Paris, 1946.

2. E. Cartan, Oeuvres Complètes, Gauthier-Villars, 1952.

1. C. Chevalley, Lie Groups, Princeton University Press, 1946.

1. R. Courant and D. Hilbert, Methods of Mathematical Physics, Interscience, New York, 1953-1962.

1. G. Darboux, Theorié Genérale des Surfaces, Chelsea Publishing Company, New York.

1. L. Ehrenpreis, Fourier Analysis in Several Complex Variables, Wiley, New York, 1920.

1. F. Estabrook and H. Wahlquist, "Prolongation Structures of Non-Linear and Evolution Equations", J. Math. Phys. 16, 1 (1975).

1. A. Forsyth, Theory of Differential Equations, Dover, New York.

1. E. Goursat, A Course in Mathematical Analysis, Dover, New York.

1. H. Goldstein, Classical Mechanics, Addison-Wesley, 1951.

1. R. Hermann, "E. Cartan's Geometric Theory of Partial Differential Equations", Advances in Math. 1, 265-317 (1965).

2. R. Hermann, Phys. Rev. Lett. 36, 835 (1976).

1. L. Hörmander, Linear Partial Differential Operators, Springer-Verlay, 1963.

1. I. Kaplansky, An Introduction to Differential Algebra, Hermann, Paris, 1957.

BIBLIOGRAPHY

1. E. Kolchin, Differential Algebra, Academic Press, New York, 1973.

1. A. Kumpera and D.C. Spencer, Lie Equations, Princeton University Press, 1972.

1. S. Lie's 1884 Differential Invariant Paper, comments and additional material by R. Hermann, Math Sci Press, Brookline, Ma., 1975.

1. H. Morris, "A Prolongation Structure for the AKNS System and its Generalization", Trinity College, Dublin, preprint, 1976.

1. R.M. Muira, C.S. Gardner, and M.D. Kruskal, "Korteweg-de Vries Equation and Generalizations", J. Math. Phys. 9, 1204-1209 (1968).

1. E. Picard, Traité d'Analyse, Gauthier-Villars, Paris, 1922.

1. C. Riquier, Les Systémes D'Equations aux Deriveés Partielles, Gauthier-Villars, Paris, 1910.

1. I. Stewart, Galois Theory, Chapman and Hall, London, 1973.

1. A. Tresse, Acta Math. 18 (1894).

1. E. Vessiot, "Etudes des Equations Différentielles Ordinaires au Point de Vue Formel", Encyclopédie des Sciences Mathématiques, Tome II, Vol. 3, Gauthier-Villars, Paris, 1910.

1. E. von Weber, G. Floquet, and E. Goursat, "Proprietés Générales des Systems d'Equations aux Derivées Partielle", Encyclopédie des Sciences Mathématiques, Tome II, Vol. 4, Gauthier-Villais, Paris, 1913.

1. G. Whitham, Linear and Nonlinear Waves, Wiley, New York, 1974.

1. E. Wilczynski, Projective Differential Geometry of Curves and Ruled Surfaces, Chelsea Publishing Co., New York.

FINAL REMARKS

(August 1976)

A GENERAL CONNECTION VECTOR BUNDLE THEORETIC FRAMEWORK
FOR THE INVERSE SCATTERING METHOD

In June, 1976 there was a conference at Ames Research Center (NASA) at which a group of us interested in the geometric side of soliton theory met, really for the first time in a unified way. Some of the material in the last chapters of this volume was written as a result of ideas developed there, and much more is to come. It seems to me even more strongly that the theory of connection is the unifying geometric framework that people have been searching for ever since the discovery of the soliton phenomena. In particular, it certainly is the key to the relation between the linear "inverse scattering" equations and the nonlinear wave-like evolution equations.

I recently came on a paper, "Relationships among Inverse Method, Bäcklund Transformation and an Infinite Number of Conservation Laws", by M. Wadati, H. Sanuki and K. Konno in Progress of Theoretical Physics, 53 (1975), pp. 419-436, which puts much of the work in an excellent perspective, and lends itself very readily to the "connection" interpretation. Here is their general framework.

FINAL REMARKS

Consider a system of partial differential equations of the following form:

$$\boxed{\begin{aligned} \frac{\partial \psi}{\partial x} &= A(\quad)\psi \\ \\ \frac{\partial \psi}{\partial t} &= B(\quad)\psi \end{aligned}} \quad (1)$$

Here, $\psi(x,t)$ is a <u>vector-valued</u> function of two real variables x and t. (Physically, x is the "space", t is the "time" variable. In the examples treated by Wadati, Sanuki and Konno ψ has two components

$$\begin{pmatrix} \psi_1 \\ \psi_2 \end{pmatrix} \quad .)$$

A, B are <u>matrix-valued functions</u> of real variables labelled z,

$$z, z_x, z_{xx}, \ldots$$

To "solve" Equation (1) means <u>to find</u> z <u>as a function</u> $z(x,t)$ <u>and</u> ψ <u>as a function</u> $\psi(x,t)$ <u>such that</u>:

$$\boxed{\begin{aligned} \frac{\partial \psi}{\partial x} &= A(z(x,t), \partial_x z(x,t), \ldots)\psi \\ \\ \frac{\partial \psi}{\partial t} &= B(z(x,t), \partial_x z(x,t), \partial_x^2 z, \ldots)\psi \end{aligned}} \quad (2)$$

FINAL REMARKS

This is to be done in the following way: Choose $z(x,t)$ so that Equations (2) are completely integrable (in the Frobenius sense). These integrability conditions require that $z(x,t)$ satisfy a certain system of partial differential equations. These latter equations for z are the "nonlinear wave equations". In the paper cited above, it is shown how to obtain the Big Three Nonlinear Wave Equations: Korteweg-de Vries, Sine-Gordon, and Modified Korteweg-de Vries by choosing A and B in an appropriate way.

Remark. It is important to keep in mind that this use of the term "complete integrability" (which is the correct one) is quite distinct from the use of the term by most of the writers on nonlinear waves. They usually mean by this the interpretation of the nonlinear wave equation as an infinite dimensional mechanical system, along with complete solution of the corresponding Hamilton-Jacobi equation. Ultimately, I am sure there will be a relation between the two ideas, but this is not at all routine or obvious, and it is completely misleading to use the term in the latter sense.

While this is clear enough as it stands, I believe it is quite significant to interpret it within the context of the theory of connections. Again, we will be able to interpret "complete integrability" as flatness of a connection, thus establishing the key geometric link between this material,

the <u>pseudo-potentials of Estabrook and Wahlquist</u>, and Bäcklund
transformation. How to do this should be obvious. Interpret

$$(x, t, z, z_x, z_{xx}, \ldots)$$

as coordinates of a manifold Y. Consider ψ as elements of
a <u>vector bundle</u> over Y, with R^n (n = number of components
in ψ; for the traditional examples n = 2) as fiber. (In
order to interpret (2) directly, it suffices to consider this
bundle as the <u>flat bundle</u> $Y \times R^n$. It might be interesting
later on to consider more exotic global possibilities.) We
can then associate (1) and (2) with the <u>following connection
forming</u> for this bundle

$$\theta = d\psi - A\psi dx - B\psi dt \qquad (3)$$

(This is, of course, <u>a vector-valued differential form</u>.)
The vanishing of its <u>curvature</u> determines "complete integrability", and determines the <u>differential equations satisfied</u>
by (3). These are readily determined by exterior differentiation:

$$-d\theta = d(A)\psi \wedge dx + Ad\psi \wedge dt + dB\psi \wedge dt + Bd\psi \wedge dt$$

$$= dA\psi \wedge dx + A(A\psi dx + B\psi dt) \wedge dx + dB\psi \wedge dt$$
$$+ B(A\psi dx + B\psi dt) \wedge dt + \cdots$$

$$= dA\psi \wedge dx + AB\psi dt \wedge dx + dB\psi \wedge dt + BA\psi dx \wedge dt + \cdots$$

FINAL REMARKS

$$= dA\psi \wedge dx + dB\psi \wedge dt + [A,B](\psi)dt \wedge dx + \cdots$$

Set:

$$\Omega = dA \wedge dx + dB \wedge dt + [A,B]dt \wedge dx \qquad (4)$$

It is the <u>curvature form</u> of the connection. Setting it equal to zero, then constraining the variables $z_x, z_{xx}, \ldots$ via the exterior relations

$$(dz - z_x dx) \wedge dt = 0$$
$$(dz_x - z_{xx} dx) \wedge dt = 0 \qquad (5)$$

determines the nonlinear wave equation.

This suggests a very general--and probably significant--setting. Let

$$\pi: E \to Y$$

be a vector bundle. Suppose given the following data:

a) A linear connection for the vector bundle whose curvature form is Ω, a vector-valued differential form on Y.

b) A differential ideal I of forms on Y.

Let ED be the <u>exterior differential system generated by</u> ED and I. The <u>nonlinear</u> waves are two-dimensional solution submanifolds of ED for which (x,t) are <u>independent variables</u>, i.e.,

$$dx \wedge dt \neq 0 \quad .$$

Cartan discusses such systems in detail in his book "Les systemes differentielles exterieure et leurs applications geométriques". It is natural to <u>conjecture</u> that the interesting choices should be <u>determined by</u> the condition that ED is <u>in involution in Cartan's sense</u>.

Hedley Morris of Trinity College, Dublin talked at our Ames Conference (his talk should be published in the Proceedings via Math Sci Press) on ways to extend the Estabrook-Wahlquist formation to equations <u>in more than two variables</u>. It will obviously be of great interest to <u>think about the geometric interpretations of this process</u>.

I hope the reader shares my enthusiasm that this subject is an extremely attractive area. Indeed, it is "applied mathematics" of the best sort; the problems are clearly related to a wide spectrum of physics and engineering (and possibly even biology) problems, and yet involve mathematical ideas sophisticated and interesting enough to involve contemporary mathematics of the highest level.

FINAL REMARKS

Finally, notice that there is an interesting "control-theoretic" aspect to this situation. Equations (2) may be thought of as a ("distributed parameter") control problem, with state vector ψ, and with z as input. I do not know if the standard control notions (controllability, equivalence, etc.) are relevant or useful, but it might be an interesting guide to further research to think of them in this way!